STATE OF THE WORLD'S FORESTS

2005

FOOD AND AGRICULTURE ORGANIZATION OF THE UNITED NATIONS

Rome, 2005

Produced by the
Editorial Production and Design Group
Publishing Management Service
FAO

The designations employed and the presentation of material in this information product do not imply the expression of any opinion whatsoever on the part of the Food and Agriculture Organization of the United Nations concerning the legal or development status of any country, territory, city or area or of its authorities, or concerning the delimitation of its frontiers or boundaries. The designations employed and the presentation of material in the map(s) do not imply the expression of any opinion whatsoever on the part of FAO concerning the legal or constitutional status of any country, territory or sea area, or concerning the delimitation of frontiers.

ISBN 92-5-105187-9

Contents

_____ PART II _____

SELECTED CURRENT ISSUES IN THE FOREST SECTOR

Foreword

S tate of the World's Forests presents a global picture of the forest sector by providing the latest information on major policy developments and key emerging issues. As with previous publications, this sixth edition aims to assist forest practitioners, other resource managers, policy experts, educators, forest industry and civil society to make informed decisions about the best way to achieve sustainable forest management.

Some people might be of the opinion that not much changes from one year to the next. Yet a look back even over the short term can leave quite another impression. Since the last publication of *State of the World's Forests*, for example, the outcomes of the World Summit on Sustainable Development are being implemented, many through partnerships. Although some participants were disappointed that forest issues *per se* were not discussed during the summit, the Plan of Implementation recognizes sustainable forest management as essential to achieving sustainable development. Note was also made of the importance of forests in eradicating poverty, improving food security and halting the loss of biological diversity. Along similar lines, many efforts to make the United Nations Millennium Development Goals a reality take into account the range of benefits that forests provide. Another milestone is fast approaching as countries participating in the United Nations Forum on Forests (UNFF) meet in May 2005 to decide on future international arrangements for forests after more than ten years of policy dialogue.

Building on the broad participatory process to assemble *State of the World's Forests 2005*, we have asked for contributions from key non-governmental organizations and from individuals acting in their personal capacity, in addition to pieces researched and written by FAO staff. The theme of this year's edition – "realizing the economic benefits from forests" – reminds us that managing forests in a sustainable manner for the range of values they provide will not be possible if the sector as a whole is not economically viable. Although markets for environmental services are growing, we know that wood and wood products will remain important sources of income for the immediate future. Therefore, governments and other owners of the resource must capture more of the production potential and create conditions for the development of efficient markets. At the same time, they must safeguard the environmental, social and cultural functions of forests.

Evidence is mounting that certain segments of society are able and willing to pay to protect watersheds, for example, and this practice is expected to become more widespread when the linkage between upstream owners and downstream users is formalized. Regulatory frameworks are also being developed to address carbon sequestration and the conservation of biological diversity. The fact remains, however, that a significant portion of forest goods and services falls outside the marketplace, causing forest owners and managers to underinvest in forest protection or sustainable forest management. Until markets for these products develop, governments will be called upon to respond to demands through public interventions or programmes.

Part I presents recent developments and areas of current attention in forest resources; the management, conservation and sustainable development of forests; the institutional framework; and the international forest policy dialogue. Part II contains five chapters, each addressing a particular subject in more detail. The first focuses on enhancing economic benefits from forests in the context of changing opportunities and challenges – a topic of interest to governments, the private sector, communities, individual forest owners and others who depend on the resource for their livelihood. The second chapter notes ways in which various agroforestry practices can benefit farmers and outlines conditions that need to be in place to maximize returns. An overview of the importance of wood energy, with a description of the economic forces that affect its production and consumption, is provided in the third chapter. The fourth chapter examines issues related to tariffs and non-tariff measures in the trade of forest products. The link between violent conflicts and forested regions, where remoteness and inaccessibility can provide cover for lucrative illegal activities, is explored in the last chapter.

Trying to do justice to key topics within the constraints dictated by length means that coverage of some has to be general rather than exhaustive. These could perhaps be treated in more depth in future editions, or other organizations and partners could consider initiating further research in areas of particular interest to them. Indeed, FAO would welcome such efforts and could offer its assistance to the extent possible.

FAO is pleased to release *State of the World's Forests 2005* and hopes that readers find it informative.

M. Hosny El-Lakany
Assistant Director-General
FAO Forestry Department

Acknowledgements

S *tate of the World's Forests 2005* is the result of extensive collaboration among people from within and outside FAO. Special thanks go to R. McConnell, who coordinated the effort and compiled the document, working closely with authors and advisers.

Appreciation is also extended to FAO staff and consultants who provided information, wrote pieces or reviewed drafts: H. Abdel-Nour, G. Allard, L. Amariei, S. Braatz, C. Brown, J. Carle, C.M. Carneiro, F. Castañeda, A. Contreras-Hermosilla, P. Csoka, P. Durst, T. Enters, J.G. Goldammer, T. Hofer, P. Holmgren, M. Jurvelius, W. Killmann, D. Kneeland, P. Kone, M. Kuzee, J. Lejeune, M. Lobovikov, T. Lopes, D. McGuire, P. McKenzie, S. Maginnis, M. Malagnoux, E. Mansur, M. Martin, M. Morell, M. Paveri, E. Pepke, A. Perlis, C. Prins, F. Romano, J. Ross, D. Schoene, E.-H. Sène, P. Sigaud, M. Trossero, T. Vahanen, P. Vantomme, P. Vuorinen, F. Wencelius, M. Wilkie and D. Williamson.

FAO is also grateful to those who contributed chapters in Part II: C.T.S. Nair, FAO, for "Enhancing economic benefits from forests: changing opportunities and challenges"; S. Franzel, World Agroforestry Centre (ICRAF), for "Realizing the economic benefits of agroforestry: experiences, lessons and challenges"; S. Kant, University of Toronto (Canada), for "Economics of wood energy"; C. Mersmann for "Tariffs and non-tariff measures in trade of forest products"; and D. Kaimowitz, Center for International Forestry Research (CIFOR), for "Forests and war, forests and peace".

FAO appreciates the dedication and valuable advice of members of the Internal and External Advisory Committees: L. Alden-Wily, J. Ball, J. Carle, M. Coulombe, H. Gregersen, C. Holding-Anyonge, W. Jackson, D. Kaimowitz, J.-P. Lanly, J. Maini, M.-R. de Montalembert, M. Morell, E. Müller, C.T.S. Nair, C. Oliver, C. Palmberg-Lerche, M. Paveri, J.A. Prado, C. Prins, S. Razak, T. Rezende de Azevedo, S. Rose, D. Schoene, R. Seppälä, M. Simula, K. Tufuor, T. Vahanen, A. Whiteman and D. Wijewardana.

A. Perlis, E. Carpenter, B. Moore and the staff of the FAO Publishing Management Service provided editorial and production support.

Summary

S tate of the World's Forests – now in its sixth edition – presents a global picture of the forest sector, providing the latest information on activities and developments. Contributions from non-governmental organizations (NGOs), individuals in their personal capacity and FAO highlight challenges and opportunities related to some of today's key emerging issues. The theme of the 2005 edition – "realizing the economic benefits from forests" – recognizes that the economic viability of the forest sector is a prerequisite to safeguarding the environmental, social and cultural functions of the resource.

SITUATION AND DEVELOPMENTS IN THE FOREST SECTOR
Forest resources
Global Forest Resources Assessment update 2005. FAO will publish the main report of the Global Forest Resources Assessment update for 2005 (FRA 2005) in the latter part of the year. The assessment focuses on key trends and builds on the thematic elements of sustainable forest management drawn from regional and ecoregional criteria and indicators processes as a reporting framework. With recent emphasis on rural livelihoods, benefit sharing, food security and how forests contribute to achieving these goals, FAO has expanded FRA reports to include social and environmental dimensions of the resource, as well as economic aspects.

Estimating carbon stock changes in forests. Developments in international discussions on climate change may alter the scope, techniques and importance of forest inventories worldwide. *State of the World's Forests 2005* notes that all Parties to the United Nations Framework Convention on Climate Change (UNFCCC) must estimate and report carbon stock changes in their forests; that the Kyoto Protocol establishes additional rules to monitor

and account for carbon stocks; and that, under special provisions for sequestration projects of Joint Implementation or the Clean Development Mechanism (CDM) of the protocol, carbon in forestry projects must be monitored in order to realize credits.

Secondary forests in tropical regions. Although figures vary according to the definition used, degraded forests and secondary forests in tropical Africa, America and Asia covered an estimated 850 million hectares in 2002. The value of secondary forests (defined here as forests regenerating largely through natural processes after significant disturbance of the original forest vegetation) for their capacity to reduce poverty, enhance food security and provide environmental services would be better recognized if foresters and decision-makers would highlight their importance to a greater extent than is now the case.

Forests and trees in small island developing states. Forests in small island developing states (SIDS) cover an estimated 75 million hectares, or 63 percent of combined land area, but forest cover differs greatly among states. Although deforestation appears to have slowed in the past decade, the average annual rate is still high in many SIDS. The main causes include conversion of forested land for agriculture and for infrastructure such as roads, ports, housing and tourism development. On the other hand, some states registered an increase in forest cover from 1990 to 2000, mainly because of afforestation. *State of the World's Forests 2005* outlines the challenges to achieving sustainable forest management in SIDS and identifies opportunities for future development of the sector.

Asia's innovative sources of raw material for industries. Plantations of rubber, coconut,

bamboo and oil-palm as well as agricultural residues are providing new sources of raw material for industries in Asia. In Malaysia, for example, exports of rubberwood products are valued at about US$1.1 billion annually. Although commercial processing of the fibres from coconut palms is still mostly for local consumption, speciality products are finding their way into niche markets and new technologies are expanding the range of items available. In recent years, strong demand and high prices for palm oil and palm kernels to make foods, soaps and cosmetics have stimulated a boom in planting of oil-palm in Asia. Technological developments have cleared the way for using bamboo in innovative ways, such as in reconstituted panel and board products. Straw, especially wheat and rice straw, is the non-wood fibre used most widely in pulp and paper manufacturing.

International trade in non-wood forest products. *State of the World's Forests 2005* presents the latest results of an ongoing FAO study on the value, trends and flows in international trade in non-wood forest products (NWFPs). It notes the problems in collecting, compiling and analysing trade data because, for example, there is no agreement among countries, agencies or authors on terminology, and NWFPs enter the market as ingredients in composite products, making them difficult to identify. From 1992 to 2002, the value of global trade in NWFPs increased 1.5 times. Before the commercialization of NWFPs can be promoted as a strategy to alleviate poverty, a number of issues need to be carefully considered, including the equitable sharing of benefits.

Management, conservation and sustainable development of forests

Sustainable forest management and the ecosystem approach. Recent international forest discussions have focused on the extent to which sustainable forest management, as outlined by the "Forest Principles" adopted by the United Nations Conference on Environment and Development (UNCED), and the ecosystem

approach, as defined by the Convention on Biological Diversity (CBD) and as applied to forests, are similar, where they differ and how they could be integrated. A comparison of the underlying principles of the two concepts reveals few differences other than that sustainable forest management deals largely with only one kind of ecosystem – forests – whereas the ecosystem approach addresses a range of ecosystems. Integrating sustainable forest management and the ecosystem approach could lead to using the same indicators to monitor and report progress, thereby reducing the reporting burden on countries. It could also result in more coordinated policy development and planning as well as better sharing of information and experiences to improve forest practices. Rather than continue the debate, efforts should now focus on implementation, building upon best practices and tools and monitoring progress.

Forest landscape restoration. There is a growing realization that in addition to conventional approaches to the sustainable management and conservation of forests to minimize further loss of the resource, restoring degraded lands at the landscape level is also necessary to guarantee a healthy, productive and biologically rich forest estate for the long term. Since the Global Partnership on Forest Landscape Restoration was launched in March 2003, organizations and governments have been exploring this concept as a complement to the management and protection of forest resources. Although it is not a new idea, its novelty lies in addressing and balancing trade-offs at the landscape level, and its pragmatic rejection of the insistence to return modified forest landscapes to their original pristine state. Forest landscape restoration is carried out under the assumption that improving the flow of forest goods and services requires balancing livelihoods with protecting nature, and that this is best achieved within dynamic, multifunctional landscapes.

Forestry and ecotourism. Much of nature tourism and ecotourism focuses on forests –

from bird-watching to canopy walks, forest treks and wildlife viewing – and can deliver significant benefits at the local and national levels. Nature tourism and ecotourism provide an incentive to protect forests and wildlife and a means by which people can generate income without extracting resources. If managed properly, ecotourism creates employment for rural communities faced with few alternative livelihood opportunities. Recent studies indicate, however, that some ecotourism previously thought to be benign stresses wildlife, disrupts breeding patterns and changes the behaviour of wild animals. *State of the World's Forests 2005* outlines some of the environmental, economic, social and cultural aspects of the industry and suggests that the recent boom will provide new challenges and opportunities for sustainable forest management around the world.

Biosecurity and invasive forest tree species. Concern over the potentially negative impact of the introduction of new species, breeding and genetic modification has increased attention on the need to develop regulatory frameworks and policies to manage risks. Introduced forest tree species can help sustain national and local economies and be of significant value to the environment and to society. However, when insufficient consideration is given prior to use and when on-site management is neglected, some species may invade adjacent areas, giving rise to a number of problems. Moreover, with global trade increasing, greater movement of people and overstretched quarantine services, the number of accidental introductions is expected to grow. Reliable information and better knowledge of economic and environmental effects are critically important for evaluating risks.

Biotechnology in forestry. Most public research in forest biotechnology is on the biology and diversity of forest tree species, populations and individuals or on propagation, rather than on genetic modification. More than two-thirds of activities on genetic diversity and marker-assisted selection are carried out in Europe and

North America, while 38 percent of research programmes using advanced propagation technology are in Asia. Most research on genetic modification in forest trees takes place in developed countries. While the tools for genetic modification in forestry are mostly the same as those in agriculture, perceptions and applications differ where forest trees are concerned because of the social, cultural and environmental aspects of forests and the fact that forest trees have only recently been domesticated, in contrast to most agricultural crop species. To improve information, FAO is now carrying out the first global review on biotechnology in forestry.

Wildland fires. Uncontrolled fires in forests, other wooded lands and other lands – generally referred to as wildland fires – continue to claim lives, destroy valuable assets and emit compounds that affect the composition and functioning of the atmosphere. Between 300 and 400 million hectares burn annually worldwide, much of it in Africa. Although the responsibility to suppress fires resides with countries and national fire authorities, the key to dealing more effectively with emergencies lies in putting agreements into place between and among states. To enhance this type of collaboration, FAO and partners are working with countries to develop bilateral or multilateral instruments.

Institutional issues

Trends in privatization in the forest sector. Governments often use privatization measures to improve economic performance, especially since the end of the 1970s. Forests, however, were not among the first assets to be privatized, partly because of the sensitivities surrounding sovereignty, a growing recognition of their importance in protecting the environment and in providing services to society, and perceived high risks or low returns. Since the 1990s, water, land and forests have become more frequent targets for privatization. This trend is less marked for natural forests than for planted forests, except in Central and Eastern Europe, where forest land is being returned to former owners. In addition, private entities and NGOs are increasingly purchasing forest areas and acquiring land through concession contracts for protection and conservation purposes. *State of the World's Forests 2005* describes the latest trends in privatization of forest resources.

Trends in forestry administration. Responding to public demand for greater accountability, more participatory decision-making and better delivery of goods and services, central forestry administrations are delegating more functions to local government. Modern reforms are changing the ways in which forests and other natural resource sectors are managed, increasing the urgency to establish partnerships, share information and coordinate activities. New technologies such as satellite imagery and detection, as well as spatial information and decision-support systems, are improving how administrations operate. In the process, staff must be taught to deal with new realities and to master emerging technologies. Steps must also be taken to ensure that all levels of authority have access to the knowledge and skills they require to perform their tasks.

Forest law compliance. Governments, with the help of international organizations, NGOs and the private sector, are continuing their efforts to improve law compliance in the forest sector. Most initiatives are built on the premise that, although important, compliance strategies can no longer rely on policing alone but must include efforts to streamline policy and legal frameworks; to provide incentives to comply with regulations; to improve employment conditions of enforcement officers; to conduct public education and awareness programmes; and to use national and international market measures to limit opportunities for trading illegally sourced wood. *State of the World's Forests 2005* describes major undertakings to date.

Forests and the Kyoto Protocol. Rules under which developed countries must measure and report their use of forests and wood products to

meet commitments to mitigate climate change under UNFCCC and the Kyoto Protocol are complicated and costly to administer. Between now and 2008 – the start of the first commitment period – countries face three major tasks with regard to implementation: putting general commitments into practice; monitoring and reporting forest carbon stock changes; and translating global commitments to mitigate climate change into law after entry into force of the Kyoto Protocol. *State of the World's Forests 2005* delves into core issues such as who owns the carbon in forests, trees and wood products.

International forest policy dialogue

Countries have been discussing international forest policy issues within the United Nations system since the end of the Second World War. Since then, the forest sector has undergone many changes. More recently, there is better recognition of the contributions that forests make to sustainable development; improved cooperation on a range of complex issues; and more participation of civil society in decision-making. However, the growing number of calls to enhance efforts to achieve sustainable forest management is overwhelming implementing agencies and many developing countries. Governments are also concerned with the number and duplication of requests for reporting to international processes. Despite the positive developments, deforestation and forest degradation continue, and illegal forest activities remain problematic, making it imperative for forest practitioners and policy-makers to reach out to other sectors to find lasting solutions. Any future international dialogue on forests should establish a broader base of experts on which to draw, including those in agriculture, infrastructure development and the energy, mining and transportation sectors. Some 13 years after UNCED, countries must either decide to give the United Nations Forum on Forests (UNFF) process a new mandate and working modalities or decide that the Ad Hoc Intergovernmental Panel on Forests (IPF) / Intergovernmental Forum on Forests (IFF) / UNFF dialogue has yielded all it can and that it

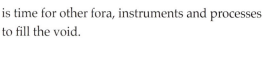

is time for other fora, instruments and processes to fill the void.

XII World Forestry Congress. In co-sponsorship with FAO, the Government of Canada hosted and organized the XII World Forestry Congress in Québec City in September 2003. Some 4 000 participants from approximately 140 countries considered topics under the theme "Forests, source of life", which was divided into three areas: forests for people; forests for the planet; and people and forests in harmony. *State of the World's Forests 2005* outlines the key outcome of the congress – a Final Statement that contains a vision, strategies and actions to achieve sustainable forest management worldwide. It calls on countries and organizations to pursue the objectives stated therein and promote them in other sectors.

SELECTED CURRENT ISSUES IN THE FOREST SECTOR
Enhancing economic benefits from forests: changing opportunities and challenges
Awareness of the economic, social, cultural and environmental contributions of forests and forestry has risen considerably in recent years, yet low investment and low incomes continue to plague the sector. Given its relatively small share of employment and national income, decision-makers give forestry a low priority in the face of competing demands for limited budgets. In response, attempts are being made to assess the value of all products and services, especially those pertaining to the environment. Efforts are also being made to develop innovative financing mechanisms and to create markets for services in order to enhance income and encourage investment in sustainable forest management. *State of the World's Forests 2005* describes ways in which communities, governments and the private sector are enhancing economic benefits from forests. It also identifies issues that must be addressed to make sustainable forest management economically viable.

Realizing the economic benefits of agroforestry
Cultivating trees in combination with crops and livestock is an ancient practice, but several factors have contributed to a growing interest in agroforestry since the 1970s: the deteriorating economic situation in many parts of the developing world; increased tropical deforestation; degradation and scarcity of land caused by population pressures; and growing interest in farming systems, intercropping and the environment. *State of the World's Forests 2005* outlines the advantages of using various agroforestry practices, describes some of the benefits to farmers and society and identifies factors that affect performance. It notes that more research is needed to quantify returns fully, to promote its wider use and to assess the effects and trade-offs of different policies. Determining which practices are most suited to women and poor people needs greater attention, as does finding ways to replicate successful interventions on a larger scale to reach more households.

Economics of wood energy
In the past decade, policies to encourage the use of renewable energy have become more important to help reduce dependence on non-renewable energy sources such as fossil fuels and as part of strategies to address global warming. Wood energy remains the most important source of energy for more than two billion people in developing countries. Wood energy is also likely to gain in popularity in developed countries over the next 20 years as part of efforts to promote the use of renewable energy. *State of the World's Forests 2005* identifies key considerations for the development of future programmes and policies, including the need to take into account the complex economic forces that influence wood energy consumption and production. In addition, it describes how countries might develop the wood energy sector to meet broad policy goals and objectives.

Tariffs and non-tariff measures in trade of forest products
Concerns over forest degradation and loss of forest cover are heightening pressure on governments, the private sector and

international institutions to address the impact and interaction between trade and the environment, and specifically their relation to sustainable forest management. Although global trade in forest products is expanding, it is influenced by trade measures that vary considerably by product, region and country, including import tariffs, export restrictions, technical product standards, sanitary and phytosanitary measures and environmental and social standards – for example, certification and product labelling. Recent international discussions have noted that trade can have both a positive and negative impact on sustainable forest management and thus have recommended that countries monitor the effects of trade policies more closely. In attempting to diversify their forest products, developing countries and countries with economies in transition need to identify national incentives, drawing upon successful experiences elsewhere in developing domestic policies yet complying with trade rules at the same time. Schemes related to certification of forest management and to labelling of forest products are improving the interaction between trade and forest management, even though complaints continue over market access and market shares, particularly of forest products from tropical regions. Trade measures are being changed and adjusted to respond to specific production and market situations, with most staying within the boundaries of global and regional trade agreements. Those that stem from concerns over sustainability in the forest sector will continue to be evaluated against special trade obligations in multilateral environmental agreements and against global and regional trade rules.

Forests and war, forests and peace
Recognizing major clashes that have taken place in Africa, Latin America and South and Southeast Asia, *State of the World's Forests 2005* examines why many violent conflicts occur in forested regions. It identifies the characteristics of recent armed disputes, looks at the links to forests, explores issues related to post-conflict situations and presents a strategy for action.

Forests offer secluded places where insurgents can hide and use valuable natural resources to finance their activities. Rebels may also engage in lucrative illegal activities such as cultivating illicit crops and smuggling. People may use violence to gain control over natural resources or because they feel neglected or mistreated. Often, reasons shift over time and combine political, religious or ethnic aspects with personal incentives such as a desire for income, wealth, status, revenge or security or loyalty to specific individuals. Efforts to promote peace in forested regions must start with removing the motives for conflict before it breaks out. Armed hostilities can have both negative and positive effects on forests. However, post-conflict situations in countries with significant forests almost always pose an acute danger for this resource. Peace requires investment in better governance and improvement of livelihoods in remote forested and mountainous regions to prevent them from serving as breeding grounds for violence. Only then can forests assume their rightful importance for the social, cultural, economic and environmental contributions they make to the lives of all who depend on them. ◆

PART I

SITUATION AND DEVELOPMENTS IN THE FOREST SECTOR

Forest resources

The last Global Forest Resources Assessment (FRA) was conducted in 2000 (FAO, 2001), and the next comprehensive assessment is expected in 2010. In line with previous interim assessments in 1995 and 1988, an update is under way for 2005 (FRA 2005) and is expected to be released later in the year. This chapter highlights the structure of the main report of FRA 2005, noting that independent studies on key global issues related to the extent and condition of forest resources will be included. The chapter also outlines reporting requirements under the United Nations Framework Convention on Climate Change (UNFCCC) and the Kyoto Protocol; underscores the importance of secondary forests in tropical regions; describes the challenges and opportunities associated with sustainable forest management in small island developing states (SIDS); provides an overview of new sources of raw material and substitutes for wood fibre in Asia; and presents the latest results of an ongoing FAO study on international trade in non-wood forest products (NWFPs).

GLOBAL FOREST RESOURCES ASSESSMENT UPDATE 2005
FRA 2005 focuses on key trends and builds on the thematic elements of sustainable forest management, drawn from regional and ecoregional criteria and indicators processes, as a reporting framework (see Box on page 3). Thus, the information collated in the assessment is relevant to national monitoring of progress towards sustainable forest management and reporting to various forest-related international organizations and processes.

FRA 2005 continues the FAO tradition of reporting on the world's forests. The periodic global assessment reports have tracked and reflected the changes that both the resource and forestry have undergone over the past 50 years. For example, for decades following the Second World War, timber supply dominated international forestry issues. Consequently, global assessments focused on the capacity of forests to yield sufficient amounts of wood in a sustainable manner. As issues related to development and the environment emerged, FRA 1980 was the first to report on deforestation and forest degradation. In 1992, the United Nations Conference on Environment and Development (UNCED) outcomes added biological diversity, climate change and desertification to the agenda. With more recent emphasis on rural livelihoods, benefit sharing, food security and how forests contribute to achieving these goals, FAO has expanded FRA reports to include social and environmental dimensions of the resource.

As coverage evolved and grew, so did the extent to which countries participated in the process. In the largest FRA gathering ever, national correspondents from 120 countries met in Rome in November 2003 to discuss issues related to the Global Forest Resources Assessment and to finalize the design of FRA 2005. Regional meetings of focal points were held throughout 2004 to support national input to global statistical tables, using agreed terms and definitions. This type of partnership has helped to make FRA widely known and accepted. In addition to enhancing the transparency of the process, regular communication and targeted assistance facilitate the documentation of methodologies and the processing of data based on information from official national sources. As a result, the FRA reports are widely acknowledged as providing the most accurate global estimates available.

The core of FRA 2005 is a set of 15 tables that are related to the thematic elements of sustainable forest management, with common terms and definitions for all countries to use (Table 1). Requests for data covering 1990, 2000 and 2005 focus on trends rather than status. The

TABLE 1

National reporting tables in FRA 2005 and links to common thematic elements of sustainable forest management

National reporting table	Extent of forest resources	Forest health and vitality	Biological diversity	Productive functions of forest resources	Protective functions of forest resources	Socio-economic functions
Extent of forest	■		■	■		
Forest ownership	■					■
Designated functions of forest			■	■	■	■
Forest characteristics	■	■	■		■	■
Growing stock	■		■	■		■
Biomass stock	■		■	■		■
Carbon stock	■			■		■
Disturbances affecting health and vitality	■	■		■	■	■
Diversity of tree species	■		■	■		■
Growing stock composition	■		■	■		■
Wood removal	■			■		■
Value of wood removal				■		■
Removal of non-wood forest products	■		■	■		■
Value of non-wood forest products				■		■
Employment in forestry						■

Note: "Forest" refers to forest and other wooded land.

exercise provides an opportunity to update data reported for 1990 and 2000 and to extend the time series to determine possible recent shifts.

Many countries are voicing concerns over the number and complexity of requests for forest-related information from international processes. They have asked for greater harmonization of efforts and for a reduction in the reporting burden. FRA 2005 has taken these concerns into account. For example, requests for data on forest biomass and carbon are consistent with information required by UNFCCC; information on threatened species is based on World Conservation Union (IUCN) classifications; employment data draw from the International Labour Organization (ILO) definitions; and information on removals is linked to reporting on forest products and trade.

While the country information in the 15 tables provides the basis for global and regional trend analysis, these tables alone cannot fully describe national status and trends in forestry because

of varying ecological, social and economic conditions. For this reason, FRA 2005 encourages countries to provide additional information on each of the common thematic elements of sustainable forest management through optional reporting. Many countries already prepare such reports for national purposes, and many developing countries use the opportunity provided by this request to work on broader national reports on sustainable forest management within the framework of the global assessment.

For each country report, documentation and background data pertinent to estimates will be archived as working papers for future reference. In addition, FRA 2005 will contain independent studies on key global issues related to the extent and condition of forest resources, including forests and water, planted forests, mangroves and forest fires.

The process of compiling a national report is an opportunity to collate information on

Criteria and indicators for sustainable forest management

The usefulness of criteria and indicators as tools to monitor and assess forest conditions and trends is recognized worldwide. They continue to increase understanding of sustainable forest management by generating better information; improve the development and implementation of forest policies, programmes and practices; strengthen stakeholder involvement in decision-making; and enhance collaboration on forest issues at the local, national, regional and international levels.

Nearly 150 countries, containing 97.5 percent of the world's forest area (FAO, 2003a), are participating in nine regional and international criteria and indicators processes.[1] As might be expected with such extensive coverage, the degree of implementation varies considerably among processes and among member countries within them.

The International Conference on the Contribution of Criteria and Indicators for Sustainable Forest Management: the Way Forward (CICI 2003) took place in Guatemala City, Guatemala, in February 2003 (FAO, 2003b). Experts underscored the contribution of sustainable forest management towards wider sustainable development and highlighted the importance of criteria and indicators in monitoring and measuring progress in achieving associated goals over time.

Drawing on the criteria of the nine processes, CICI 2003 acknowledged that sustainable forest management comprises seven common thematic elements:
- extent of forest resources;
- biological diversity;
- forest health and vitality;

- productive functions of forest resources;
- protective functions of forest resources;
- socio-economic functions;
- legal, policy and institutional framework.

In March 2003, the sixteenth session of the FAO Committee on Forestry (COFO) took note of this development, and less than one year later the FAO/International Tropical Timber Organization (ITTO) Expert Consultation on Criteria and Indicators for Sustainable Forest Management, held in Cebu City, Philippines, recognized the potential for these elements to facilitate communication on forest issues internationally. Delegates at the fourth session of the United Nations Forum on Forests (UNFF) in May 2004 also acknowledged that the seven elements offer a reference framework for sustainable forest management (see page 58).

From a practical perspective, FRA 2005 is building on the common thematic elements of sustainable forest management as a reporting framework, and the Collaborative Partnership on Forests (CPF) is using them as a basis for developing an information framework for forest reporting (see page 59).

---

[1] The African Timber Organization (ATO) Process, the Dry Forest in Asia Process, the Dry-Zone Africa Process, the International Tropical Timber Organization (ITTO) Process, the Lepaterique Process of Central America, the Montreal Process, the Near East Process, the Pan-European Forest Process and the Tarapoto Proposal for the Sustainability of the Amazon Forest.

several important indicators for better decision-making and to illustrate progress towards sustainable forest management at the national level. Besides a number of other uses, the periodic FRA reports serve as references to those involved in national forest programmes, forest

outlook studies and the preparation of reports to intergovernmental processes on forests and sustainable development. Finally, FRA findings form a significant input into reports on progress in achieving the United Nations Millennium Development Goals.

Further information on FRA is available at www.fao.org/forestry/fra.

NEW GUIDANCE FOR ESTIMATING CARBON STOCK CHANGES IN FORESTS

The scope, techniques and importance of forest inventories worldwide may change as a result of developments in the international dialogue on climate change. First, all Parties to UNFCCC must estimate and report carbon stock changes in their forests under the convention's rules. Developed countries report annually, developing nations periodically. Second, the Kyoto Protocol establishes additional rules to monitor and account for carbon stocks. Developed countries (and the European Community) that have ratified the protocol must supplement their reporting under UNFCCC with disaggregated and more precise estimates for each year of the commitment period. By the end of 2006, these countries must also put in place an inventory system to record carbon removals and emissions, including those from forests. Finally, under the special provisions for sequestration projects of Joint Implementation or the Clean Development Mechanism (CDM) of the Kyoto Protocol, carbon in forestry projects must be monitored in order to realize credits. Thus, developed countries wishing to offset national emissions with carbon credits gained in their forests must periodically measure forest carbon to benefit from its market value. Partners involved in forestry sequestration projects must do likewise.

After two years of work involving about 120 experts, the Intergovernmental Panel on Climate Change (IPCC) delivered its Good Practice Guidance for Land Use, Land-Use Change and Forestry (GPG). The report (IPCC, 2004) defines inventory and calculation methods that reduce uncertainties as far as possible and neither overestimate nor underestimate carbon stock changes. Although Parties have so far hedged on whether and how to account for carbon in harvested wood products, the GPG outlines methods for assessing this potentially large carbon store as well.

The GPG combines two basic methods for estimating stock changes in the carbon pools of forest ecosystems and uses a progressively sophisticated three-tier system for calculations. Tier 1 acknowledges the lack of specific national data in many countries and employs simple methods, aggregated figures (e.g. one deforestation rate that covers all forest ecosystems in a country) and rough default values (e.g. the average carbon sequestration rate per hectare of all forests). Such values are offered in appendixes of the GPG and are often based on FAO statistics. Tier 3 uses country-specific data and less aggregated activities and may use computer modelling. Tier 2 is a combination of Tiers 1 and 3. Under UNFCCC rules, all carbon pools – living biomass above and below ground, dead wood, litter and soil organic matter – must be assessed. However, for reasons of practicability and efficiency, the GPG allows pools to be treated with variable intensity. Carbon pools that are key contributors to emissions should be assessed with Tier 2 or 3 methods, while Tier 1 suffices for less significant categories.

Under stricter rules for carbon monitoring in the Kyoto Protocol, developed countries may omit a certain carbon pool from national accounting after providing transparent and verifiable information that it is not a source of carbon emissions. Participants in forestry projects under the CDM may also forego possible credits by choosing to disregard carbon pools that are difficult to measure, for example soil or dead wood, as long as there is proof that these pools will not release carbon during the crediting period.

The GPG prescribes two basic methods for assessing carbon stock changes – the default method and the stock change method – each requiring more effort, resources and data and increasing in reliability the higher the tier chosen.

The default method estimates carbon change based on the difference between periodic carbon gains and periodic carbon losses. Gain is defined as the product of growing stock increment, wood density, biomass expansion factor, root–shoot ratio and biomass carbon fraction. Loss represents the sum of fellings, fuelwood gathering and natural calamities all expressed as biomass and carbon via appropriate expansion factors. The stock change method estimates carbon change based on the difference in biomass stocks between two periodic inventories where biomass, at each point in time, is the product of growing stock, wood density, biomass expansion factor, root–shoot ratio and biomass carbon fraction.

Carrying out inventories according to the GPG by either method may be a formidable task for developed countries, given that some of the data and parameters for higher tiers may not be reliable. Gaps include actual increment, losses from harvest and calamities, specific biomass expansion factors, fellings, removals and root–shoot ratios. Monitoring carbon accumulation in sequestration projects will also necessitate new knowledge, and may markedly increase transaction costs, particularly for smallholders and community forestry projects. Parties to UNFCCC have realized this and are working on simplifying rules for these smaller initiatives.

In many developing countries, where emissions from industry are relatively minor, deforestation and forest harvesting are likely to represent a key source of greenhouse gases reported under UNFCCC. In Africa, for example, land-use change – essentially deforestation – contributes approximately 70 percent of emissions. However, basic information such as forest area can be highly questionable as more than half of developing countries that reported for FRA 2000 based their inventories on estimates rather than on ground sampling or aerial surveys. Only 2 percent of countries conducted more than one national forest inventory, and not one could report forest increment. Hence, data in these countries for both the default and stock change methods are either uncertain or missing altogether, so that calculations of carbon stock changes in forests using default values are likely to have a large margin of error.

National forest resources assessments have long been recognized as essential tools for forest policy and national development. However, global climate change, obligations under UNFCCC and the Kyoto Protocol, and

the GPG add to the need to close information gaps and to improve the frequency, accuracy and quality of such assessments. For countries to benefit economically from the carbon services provided by forests and for them to comply with new reporting obligations, concerted research efforts, along with intensified and adapted forest inventory techniques, may be necessary. In this respect, the FAO programme to support national forest assessments and the process being used to update FRA help build country capacity to generate additional information to meet current and emerging needs.

SECONDARY FORESTS IN TROPICAL REGIONS

Secondary forests are defined as "forests regenerating largely through natural processes after significant disturbance (human and natural) of the original forest vegetation at a single point in time or over an extended period, and displaying a major difference in forest structure and/or canopy species composition with respect to nearby primary forests on similar sites" (FAO, 2003c).

Areas of secondary forest throughout the tropics are increasing dramatically, and in many tropical countries they now exceed areas covered by primary forest. Most of these secondary forests develop following the disturbance or elimination of natural forests by slash-and-burn practices, conversion to agricultural activities and subsequent abandonment of lands or following excessive logging operations that have reduced the original forest to a non-commercial resource. In both cases, seeds from surrounding trees have led to eventual regeneration of the forest.

Although figures vary according to the definition used, the extent of degraded forests and secondary forests in tropical Africa, America and Asia in 2002 was estimated at 245 million, 335 million and 270 million hectares, respectively, for a total 850 million hectares (ITTO, 2002). According to FAO (2001), the reported loss of natural forests in the tropics during the 1990s was approximately 15.2 million hectares annually, of which 90 percent or more was converted to other land uses. These

estimates indicate that the potential future area of secondary forests could be considerable.

To some people, the term secondary forests may imply that they are less important than primary forests. However, they provide a wide array of goods and services to society, especially to local communities that depend on this resource (see Box below). Secondary forests remain undervalued and underused for their capacity to reduce poverty, enhance food security and provide environmental services. In part, this is because foresters and decision-makers do not sufficiently highlight their importance. Lack of knowledge on how

---

Products, goods and services from secondary forests

Secondary forests:
- provide environmental services such as regulation of water flow and quality, erosion control and carbon sequestration;
- are used in agricultural systems to restore nutrients and soil properties, to prevent pests and diseases and to regulate microclimate, which may be beneficial for the establishment of certain species (e.g. shade for intercropping with coffee and cocoa);
- provide many NWFPs (e.g. medicines, fruits, grains, game and bushmeat, bamboo and rattans) which are more readily harvested from secondary forests because of their relative accessibility;
- provide wood products such as timber, construction wood, fuelwood and charcoal;
- contribute to biodiversity conservation by relieving pressure on primary forests, by functioning as corridors for the migration of flora and fauna in fragmented landscapes and by maintaining plant and animal genetic resources.

---

to manage this resource sustainably is also a constraint.

More information is needed on the status and extent of secondary forests and on management options. To obtain it, agreement must be reached on a common definition and on which forest types to include. In addition, experiences in managing this resource need to be documented, shared and widely distributed for further possible use and adaptation. Their contribution to the livelihoods of rural communities and national development goals also has to be acknowledged.

Currently, secondary forests *per se* are not prioritized or addressed in national forest programmes, national development strategies or forest inventories. This omission has often led to their undervaluation and their conversion to planted forests or other land uses such as agriculture. As in the case of primary forests, overexploitation has led to the degradation of secondary forests, making them prone to colonization by invasive species. The cost of restoring these degraded forests is high, and the process is slow as well as difficult.

For the past few years, a number of international and regional organizations such as the Tropical Agricultural Research and Higher Education Center (CATIE), the Center for International Forestry Research (CIFOR), the World Agroforestry Centre (ICRAF), the International Tropical Timber Organization (ITTO) and FAO, in collaboration with the donor community, have been raising awareness of the importance of secondary forests and are working to improve management practices. As part of these efforts, regional workshops were organized in Latin America (Peru, June 1997), Asia (Indonesia, November 1997 and April 2000) and Africa (Kenya, December 2002 and Cameroon, November 2003). Discussions highlighted lessons learned. In addition, participants drew a number of conclusions and made several recommendations.
- As with primary forests, secondary forests are a good source of wood fibre, NWFPs, social and environmental services and other goods.

- Forest policy and legislation should take into account that requirements for secondary forests differ from those of primary forests and need to be identified, highlighted and addressed separately.
- Appropriate management options that take into account the needs of people who depend on this resource for their livelihoods need to be identified and implemented.
- Successes and lessons related to the sustainable management of secondary forests need to be widely publicized. Those involved should exchange information and experiences on policy development and implementation and on social, economic, ecological and technical management aspects.
- Countries should catalogue and assess secondary forests and their typologies, and efforts should be made to recognize fully their contribution to local, regional and national economies.
- Secondary forests should feature prominently on the international political agenda, in national policies and in national forest programmes. They should be managed as an integral part of land use, and their

contributions should be highlighted in poverty reduction strategies.

FORESTS AND TREES IN SMALL ISLAND DEVELOPING STATES

In January 2005, Mauritius hosted the International Meeting for the Ten-Year Review of the Barbados Programme of Action on Small Island Developing States. Participants assessed progress in implementing the programme, reinforced commitments and further explored ways forward, including those regarding forestry issues, which formed part of the programme's chapter on land resources.

Despite the lack of an agreed definition of a small island developing state, the establishment in 1991 of the Alliance of Small Island States (AOSIS) gave SIDS an international political identity. AOSIS has 39 member countries, including four low-lying coastal states – Belize, Guinea-Bissau, Guyana and Suriname – and four dependent territories as observers. FAO statistics for SIDS also include Bahrain and the Dominican Republic, which are members of FAO but not of AOSIS, bringing the total number to 41 (Table 2).

TABLE 2
Countries considered small island developing states in FAO reporting as of September 2002

Africa
Cape Verde
Comoros
Guinea-Bissau
Mauritius
Sao Tome and Principe
Seychelles

Asia
Bahrain[a]
Cyprus
Maldives
Singapore[b]

Europe
Malta

North and Central America
Antigua and Barbuda
Bahamas
Barbados
Belize
Cuba
Dominica
Dominican Republic[a]
Grenada
Haiti

Jamaica
Saint Kitts and Nevis
Saint Lucia
Saint Vincent and the Grenadines
Trinidad and Tobago

Oceania
Cook Islands
Federated States of Micronesia
Fiji
Kiribati
Marshall Islands
Nauru
Niue
Palau
Papua New Guinea
Samoa
Solomon Islands
Tonga
Tuvalu
Vanuatu

South America
Guyana
Suriname

[a] Not a member of AOSIS.
[b] Not a member of FAO.

Forest resources

Forests in SIDS cover an estimated 75 million hectares, or 63 percent of combined land area (FAO, 2002), but the extent of forest cover differs greatly among islands. For example, the forest cover of the Bahamas, the Cook Islands, Palau, Solomon Islands and two low-lying coastal states (Guyana and Suriname) ranges from 76 to 96 percent of total land area. Conversely, 11 of the 41 SIDS (Bahrain, Barbados, the Comoros, Haiti, Maldives, Malta, the Marshall Islands, Mauritius, Nauru, Singapore and Tonga) have forest cover of less than 10 percent, and some of these have less than 1 percent. No data are available for Tuvalu. The combined forest cover of island states with a land area of less than 50 000 km² (i.e. excluding the low-lying coastal states, Cuba and Papua New Guinea) was estimated at 38.4 percent of total land area in 2000, compared with the world average of 29.6 percent.

Although deforestation appears to have slowed in the past decade, the average annual rate is still high in many SIDS. Of the ten countries with the highest annual deforestation rates between 1990 and 2000, four are SIDS (the Comoros, the Federated States of Micronesia, Haiti and Saint Lucia). The main causes include conversion of forested land for agriculture and for infrastructure such as roads, ports, housing and tourism development. On the other hand,

Bahrain, Cape Verde, Cuba, Cyprus, Grenada and Vanuatu registered an increase in forest cover from 1990 to 2000, mainly because of afforestation. Table 3 shows the change in forest cover in the 41 island states and worldwide between 1990 and 2000.

While the total forest cover represents less than 1 percent of the world's forest area, these forests and trees are essential locally because they enhance food security, in part by protecting marine and coastal environments, and regulate the quantity and quality of water supplies. In addition, forest resources on several islands are important globally in terms of biodiversity conservation. For most of the larger islands, forests also contribute significantly to the national economy.

Challenges to sustainable forest management

Small island countries vary a great deal according to geographic, ecological, political, social, cultural and economic characteristics, but many share similar constraints to the sustainable conservation and use of their forest resources:

- limited land area and high population pressure particularly in lowland and coastal areas;
- vulnerability to environmental disasters and climate change, including rising sea levels associated with global warming;

TABLE 3

Forest cover in small island developing states, by region, 1990 and 2000

Region	Total forest ('000 ha)		Annual change 1990–2000	
	1990	2000	'000 ha	%
Africa	2 524	2 353	-17	-0.70
Asia	122	175	5	3.67
Europe	n.s.	n.s.	n.s.	n.s.
North and Central America	6 902	6 667	-24	-0.35
Oceania	35 832	34 614	-122	-0.35
South America	31 478	30 992	-49	-0.16
All 41 SIDS	76 858	74 801	-206	-0.27
SIDS <50 000 km²	7 472	7 325	-15	-0.20

Note: n.s. = not significant.
Source: FAO, 2002.

FORESTRY DEPARTMENT, JAMAICA

Small island developing states face a variety of challenges to forestry and sustainable forest management, including scarcity and inaccessibility of land. In Jamaica, where scattered timber trees are a feature of farm landscapes, the Forestry Department has engaged local communities in participatory forest management and provided training in appropriate silvicultural techniques

- high species endemism and high risk for loss of biological diversity because of small population sizes;
- alien species that are difficult to control;
- small tracts of forest over vast areas and in geographic isolation resulting in high costs for public administration and infrastructure (including transport and communications), small internal markets, limited export volumes and reduced competitiveness;
- weak institutional capacity;
- insecure land tenure and absentee landowners;
- lack of integrated land-use planning.

Opportunities and future prospects

Although many of the larger SIDS are well endowed with forests, not all forests are accessible, and harvesting of commercial species has already reached unsustainable levels in many places. Future increases in wood production will depend on more countries' adopting sound harvesting practices and applying appropriate silvicultural techniques. Additional production from planted forests is possible in larger SIDS, but scarce land restricts the potential for large-scale operations in many of these countries. Lack of good soils is also a problem, particularly in SIDS that are coral based. Agroforestry systems with coconut as the main wood-like resource seem to hold the most promise for atolls with low soil fertility and for smaller islands where land is limited.

Value-added wood processing, in particular of local hardwoods, offers good prospects for diversification in SIDS that are well endowed with forests. Diversification is also possible in terms of NWFPs where niche markets are present or can be developed, and in terms of bioprospecting, given that many SIDS have unique genetic resources.

Tourism is a key industry in many SIDS and, with interest in ecotourism and nature tourism on the rise (see page 27), forests may well contribute to growth in the sector. Efforts to develop the industry will need to be made in the context of an integrated plan that takes into account social, cultural and environmental dimensions.

Considerable potential also exists to develop markets for environmental services, coupled with mechanisms to compensate resource owners for their provision.

The varied and important roles of forests and trees in SIDS call for holistic and integrated approaches to their sustainable management, which should take into account not only the direct benefits that they provide, but also their links with other natural ecosystems and sectors such as tourism. Although SIDS are diverse and spread around the globe, they share many constraints and prospects. The extent to which they can overcome limitations and capitalize on emerging opportunities depends on political will (including at the community level), regional collaboration and international support – not least in terms of

disaster reduction strategies and assistance when calamities strike.

ASIA'S NEW WOODS AND FIBRES

Plantations of agricultural and industrial crops such as rubber, coconut, bamboo and oil-palm are providing new sources of raw materials for forest industries in Asia. In addition, agricultural residues are important substitutes for wood fibre. Several of these "new woods and fibres" are being used to manufacture traditional and new forest products in Asia.

Rubberwood

Rubberwood (*Hevea brasiliensis*) has been planted throughout Southeast Asia during the past century for the production of latex. Plantations now cover an estimated 9 million hectares, making it the most widely planted tree species in Asia (FAO, 2001) (see Table 4).

Rubberwood was introduced into markets in the late 1970s after developments in seasoning and preservative treatments improved its feasibility as a utility timber. For the past decade, it has been an important raw material in the mix of Southeast Asian wood products, particularly for those destined for export.

The viable rubberwood harvest in Southeast Asia is estimated to exceed 6.5 million cubic metres per year (Balsiger, Bahdon and Whiteman, 2000). Most is processed into sawnwood and further into furniture. As a medium-dense hardwood with light colour, easy machining and good staining properties, rubberwood can be a substitute for many species, including ramin (*Gonystylus* spp.), meranti (*Shorea* spp.), teak (*Tectona grandis*), oak (*Quercus*

spp.) and pine (*Pinus* spp.). It is increasingly being used in particle board, plywood, cement board and medium-density fibreboard, and trials are being conducted for use in oriented strandboard. Rubberwood accounts for more than 80 percent of Malaysia's furniture output, and exports of rubberwood products are valued at about US$1.1 billion. Thailand also has a large rubberwood furniture industry, with exports totalling more than US$300 million annually.

Coconut

The coconut palm (*Cocos nucifera*) is an agricultural crop throughout Asia and the South Pacific. Copra, from which coconut oil is derived, is the primary harvest. Of the more than 10 million hectares of coconut plantations in Asia, at least 2.1 million hectares are more than 60 years old, the age at which copra yields decline (Killmann, 2001).

Botanically, the coconut palm belongs to the monocotyledons, so its fibres are not classified as wood. When special processing and grading techniques are applied, parts of the stems of old coconut palms of the tall varieties can be used as a wood substitute (Killmann and Fink, 1996). At present, coconut sawing is mainly on a small scale, although volumes are significant. Timber is usually cut to meet local needs, mostly replacing timber from tree species traditionally used in rural houses and bridges. Commercial processing of coconut wood began in the 1970s in the Philippines for construction, pallets, stairs, windows and doorposts, tool handles, flooring and power poles. While coconut lumber is still mostly sold in domestic markets, specialty products such as decorative wall panelling,

TABLE 4
Areas of major woody agricultural plantation crops in Asia (*'000 ha*)

Crop	Indonesia	Malaysia	Philippines	Thailand	Others	Total
Rubberwood (1997)	3 516	1 635	88	1 966	1 705	8 910
Coconut (1997)	3 760	270	3 314	377	2 593	10 314
Oil-palm (1999)	1 807	3 313	n/a	155	35	5 310
Total	**9 083**	**5 218**	**3 402**	**2 498**	**4 333**	**24 534**

Note: n/a = not available.
Source: Killmann, 2001.

parquet and blockboard are finding their way into niche markets.

Coconut wood is also used for non-construction purposes. New technologies enable its processing into a range of products, including cabinets and crafts such as jewellery boxes, cups, vases, plates and bowls (Arancon, 1997).

Oil-palm

In recent years, strong demand and high prices for palm oil and palm kernel have stimulated a boom in planting of oil-palm (*Elaeis guineensis*) in Asia to make foods, soaps and cosmetics. Plantations in Asia covered more than 5.3 million hectares in 1999 (Killmann, 2001) (Table 4).

At the time of felling, oil-palms produce an average of 235 cubic metres of stem material per hectare. This means that approximately 50 million cubic metres of residue will be generated each year in Asia over the coming decades, depending on the rate of replanting, which is often influenced by incentives and market prices for palm oil. In addition, as much as 100 million tonnes of palm fronds, 20 million tonnes of empty fruit bunches and 5 million tonnes of palm kernel shells are produced

annually as by-products and could be available for processing.

The physical and mechanical properties of oil-palm stems, like those of coconut palm stems, vary considerably over cross-section and height. The low recovery rate and high moisture content result in considerable transport and seasoning costs (Killmann and Woon, 1990) and make oil-palm stem material uneconomical as a substitute for solid timber. However, research is progressing and trials utilizing oil-palm fibre in mechanical and chemical pulping processes show promise. Research on utilizing oil-palm in wood panels and in gypsum fibreboard has advanced faster (Kollert, Killmann and Sudin, 1994). The use of palm fronds in moulded furniture, the production of particle board and activated carbon and the sawing and lamination of palm trunks are also being investigated (Razak, 2000).

Bamboo

Material from the stems of monocotyledonous bamboo has a long history of use in Asia and from this perspective barely qualifies as a "new wood" (see also Box on page 12). However, many new uses are opening up opportunities for growers and processors. Bamboo culms (stems) have been traditional substitutes for timber in construction and scaffolding, and these uses remain vitally important in Asia today. Recent technological developments have cleared the way for using bamboo in reconstituted panel and board products (Ruiz-Pérez *et al.*, 2001). Bamboo furniture and flooring are gaining market share, with new and innovative designs contributing to this growth.

China and India have the world's largest bamboo resources (see Box on page 12). China is the world's largest producer of commercial

The many uses of bamboo – including corrugated roofing and paper making – are opening up opportunities for growers and processors in Asia

M. LOBOVIKOV

Products, goods and services from secondary forests

Found in tropical, subtropical and often temperate zones, bamboo is an ancient woody grass that numbers up to 90 genera and 1 500 species, of which only about 50 species are domesticated so far.

More than 1 billion people live in bamboo houses, and 2.5 billion people depend on this resource for their livelihoods. Besides traditional use for construction, furniture, handicrafts and food, bamboo is increasingly being recognized as an environmentally friendly and cost-effective wood substitute for producing pulp, paper, boards, panels, flooring, roofing, composites and charcoal. Bamboo shoots are rich in fibre and are competing vigorously in the international vegetable market. In addition, bamboo has significant potential to help cope with wood shortages, reduce deforestation and reverse environmental degradation. While only 10 to 20 percent of bamboo consumed reaches international markets, the value of annual trade is about US$5 to $7 billion. By comparison, trade per year in tropical timber and in bananas is estimated at about US$8 billion and $5 billion, respectively.

Although global statistics on bamboo resources are poor despite sophisticated assessment techniques developed over the past 20 years, figures are improving at the national level. India reports 9 million hectares of bamboo forest and clumps, China counts 7 million hectares of bamboo, including 4 million hectares of plantations, and Ethiopia has 2 million hectares. Latin America is presumed to have 11 million hectares of bamboo, excluding the Andes (Jiang, 2002). According to the lowest estimates, bamboo makes up about 1 percent, or 22 million hectares, of tropical and subtropical forest cover with an annual sustainable harvest of about 20 million tonnes. If current trends persist, bamboo area and stock proportion are expected to double or triple by 2025.

Headquartered in Beijing, China, the International Network for Bamboo and Rattan (INBAR), through participating organizations and individuals from all continents, develops and assists in the transfer of technologies and solutions to benefit people and their environment.

bamboo, with unprocessed bamboo valued at US$1.5 billion in 1999 (Ruiz-Pérez *et al.*, 2001). Processing is estimated to have added another US$1.3 billion to this total. The sector provides part- or full-time employment for more than 5 million people in China.

India's annual harvest totals approximately 4 million tonnes, with slightly more than half used in rural construction and scaffolding (Ganapathy, 2000). Most of the remainder is for making pulp and paper. Bamboo is also used extensively to make paper in Bangladesh, China, the Philippines, Thailand and Viet Nam.

Household fruit- and timber trees
Most countries in Asia use at least some wood from fruit-trees as commercial timber, and a

growing number of households have become important sources of this wood. In Sri Lanka, for instance, where the enforcement of a logging ban in natural forests has necessitated the use of alternative sources of timber, an estimated 500 000 cubic metres of logs (40 percent of the country's supply) come from home gardens (Bandaratillake, 2001; Ariyadasa, 2002). In the densely populated Indian State of Kerala, an estimated 83 percent of all timber (12 million cubic metres per year) comes from homesteads (FAO, 2001).

Several fruit-tree species such as jackfruit (*Artocarpus heterophyllus*) and tamarind (*Tamarindus indica*) supply high-value wood for furniture and cabinets in several countries of the region. More traditional timber species

In collaboration with the United Nations Environment Programme World Conservation Monitoring Centre (UNEP-WCMC), INBAR has developed an innovative approach to quantify and map the likely range and distribution of bamboo species (Bystriakova *et al.*, 2003; Bystriakova, Kapos and Lysenko, 2004). Figure 1 is an example.

FIGURE 1

Natural distribution and matching sites for the bamboo species *Phyllostachys pubescens*

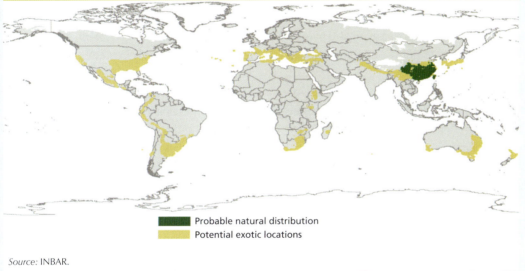

    Probable natural distribution
    Potential exotic locations

Source: INBAR.

such as mahogany (*Swietenia macrophylla*) and teak (*Tectona grandis*) and other trees such as kapok (*Ceiba pentandra*), domba (*Calophyllum inophyllum*), mango (*Mangifera indica*), durian (*Durio zibethinus*) and sapu (*Michelia champaca*) are also grown in home gardens and commonly used.

On the island of Bali, Indonesia, where carving is an important feature of local culture and livelihoods, *Paraserianthes falcataria* grown alongside rice fields has become an important raw material. In Thailand, wood from the rain tree, also known as monkey pod (*Samanea saman*), has replaced scarce teak wood in the carving industry. Rain tree wood is also increasingly used to make furniture in the Philippines, Thailand and other Asian countries.

Other fibre sources

Agricultural residues have been used in paper making in Asia for centuries, but recent technological advances in collection and handling have spurred production of non-wood pulp to more than 16 million tonnes (FAO, 2004), of which 14 million are produced in China.

Straw, the main by-product of grain harvesting, is the non-wood fibre used most widely in pulp and paper manufacturing in Asia, with wheat and rice straws the most common. Of the more than 10 million tonnes of straw-pulping capacity in Asia, China alone accounts for 9.7 million tonnes. Other major producers of straw-based papers are India, Indonesia, Pakistan and Sri Lanka.

Bagasse, the fibrous residue left after extracting juice from sugar cane, is another important fibre source. In India, the world's largest sugar producer, an estimated 7.2 million tonnes of bagasse could be used in making pulp and paper (Ganapathy, 2000). Currently, India produces more than 1 million tonnes of paper from bagasse each year and has the world's largest bagasse mill. Other countries producing this type of pulp include Bangladesh, China, Indonesia, Pakistan and Thailand. In the manufacturing of particle board, 70 percent of Pakistan's production was made from bagasse as early as the 1980s (Killmann, 1984). Reeds, jute, abaca (manila hemp) and kenaf (*Hibiscus cannabinus*) are other non-wood fibres used in making pulp and paper.

Innovative uses of agricultural residues are emerging to manufacture reconstituted panels and boards in Asia. In Malaysia, for example, rice-husk boards for flooring, panelling and furniture have created considerable interest during the past five years because of their high silica content, which renders them durable and termite proof.

Generally equivalent to medium-density fibreboard in strength and appearance, strawboard is another panel product being commercialized. Mills are being established in several Asian countries. The product has cost and environmental advantages in that disposal of straw is expensive if it is ploughed under and polluting if it is burned.

Prospects

Extrapolating the expansion rates of the main "woody" agricultural plantation crops recorded in Asia reveals an estimated 27.4 million hectares of a largely untapped resource. Meanwhile, other woody species such as bamboo and various fruit-trees, along with agricultural residues, are significant in the production of pulp, paper, reconstituted boards and specialty products. Asia's history of innovation in the forestry sector suggests that these "new woods and fibres" will have an increasingly important influence on the forest products industry in decades to come.

INTERNATIONAL TRADE IN NON-WOOD FOREST PRODUCTS

As defined by FAO, NWFPs consist of goods of biological origin other than wood, derived from forests, other wooded land and trees outside forests. This section presents latest results of an ongoing FAO study on the value, trends and flows in international trade in NWFPs over the past decade. The assessment was mainly based on a review of the *Comtrade* database (UN, 2004), which aggregates data on traded commodities as reported by national customs agencies in accord with the International Convention on the Harmonized Commodity Description and Coding System, also referred to as the Harmonized System (HS) (WCO, 2004). Where needed and possible, the information was supplemented with data from national customs agencies of major trading countries.

Tables 5 and 6 present total import values of raw materials, as well as semi-processed and processed products, for 1992 and 2002. All figures are in current rather than real US dollars, so that the growth in trade of most goods appears more than it actually is.

Most of the 28 commodities listed in Table 5 are unprocessed, although a few semi-processed products are included. For 2002, their total import value amounted to US$2.7 billion. Excluding the two commodities that were not coded in 1992 (mushroom categories 070959 and 071239), the total value of the remaining 26 increased from US$1.9 to $2.1 billion between 1992 and 2002. Eleven rose in value, eight remained the same and seven declined.

Table 6 lists 34 commodities at different stages of processing, originating from both inside and outside forests, with a total import value for 2002 of US$7 billion. By comparison, the value of global imports of wood-based forest products for the same year, including fuelwood and charcoal, amounts to US$141.4 billion (FAO, 2004). Excluding the five commodities for which trade data cannot be compared because codes did not exist in the 1992 HS, the total value of trade of the remaining 29 increased from US$4 billion in 1992 to $6.2 billion in 2002. It increased for 21, remained the same for three and declined for five.

TABLE 5
Global import values of key NWFPs for which HS code refers to a single product, 1992 and 2002

HS code	Commodity description	Global import value ('000 US$)	
		1992	2002
060410	Mosses and lichens for bouquets, ornamental purposes	9 352	25 476
070952	Truffles, fresh or chilled	4 201	23 656
070959	Mushrooms other than *Agaricus*, fresh or chilled	n.a.	364 412
071239	Mushrooms (excl. 071331/33) and truffles, dried	n.a.	219 458
200320	Truffles, prepared or preserved, not in vinegar	3 049	11 012
080120	Brazil nuts, fresh or dried	44 344	59 848
080240	Chestnuts, fresh or dried	109 958	184 663
230810	Acorns and horse chestnuts for animal feed	1 216	7 380*
120792	Sheanuts (karite nuts)	5 155	5 136*
121110	Liquorice roots	33 455	24 310
121120	Ginseng roots	389 345	221 435
121190	Plants and parts, pharmacy, perfume, insecticide use n.e.s.	689 926	777 980
121210	Locust beans, locust seeds	22 395	40 239
130110	Lac	25 286	25 653
130120	Gum arabic	101 312	105 510
130190	Natural gum, resin, gum resin, balsam, not gum arabic	92 755	96 535
400130	Balata, gutta-percha, guayule, chicle and similar gums	26 726	13 605
130214	Pyrethrum, roots containing rotenone, extracts	27 865	26 173*
140110	Bamboos used primarily for plaiting	37 562	50 054
140120	Rattan used primarily for plaiting	118 987	51 327
140210	Kapok	11 920	2 826*
170220	Maple sugar and maple syrup	43 632	116 202
200891	Palm hearts, otherwise prepared or preserved	16 082	67 514
320110	Quebracho tanning extract	51 938	45 173
320120	Wattle tanning extract	63 877	34 168
320130	Oak or chestnut extract	8 653	917*
450110	Natural cork, raw or simply prepared	7 874	110 702
530521	Abaca fibre, raw (*Musa textilis*)	15 221	20 374

* 2001 value (as no longer in HS 2002).
Notes: n.a.: not applicable as this code did not exist in the HS 1992 version.
n.e.s.: not elsewhere specified.
Source: UN, 2004.

Between 1992 and 2002, import values of the 55 commodities in the two tables increased by 50 percent, from US$5.5 billion to $8.3 billion. However, the total global import value of all commodities listed in the 1992 and 2002 HS, as recorded by trading countries, increased almost two and a half times, from US$2.24 trillion to $5.56 trillion. In addition, the share of global trade of the 55 commodities decreased from 0.25 percent to 0.15 percent, mostly as a result of a decline in the price of raw materials and because other materials gained in popularity.

Products that saw no real increase in their traded values are shea nuts, gum arabic, balata, gutta-percha, kapok, tanning extracts of quebracho and black wattle, Brazil nuts,

TABLE 6

Global import values of selected commodities for which HS code includes NWFPs among others, 1992 and 2002

HS code	Commodity description	Global import value ('000 US$)	
		1992	2002
010600	Animals, live, except farm animals	183 922	404 633
030110	Ornamental fish, live	137 886	240 965
040900	Honey, natural	268 184	657 612
041000	Edible products of animal origin n.e.s.	80 389	175 770
051000	Ambergris, civet, musk, etc. for pharmaceutical use	134 088	93 942
060491	Foliage, branches, for bouquets, etc. - fresh	n.a.	587 689
060499	Foliage, branches, for bouquets, etc. - except fresh	n.a.	103 998
071230	Mushrooms and truffles, dried, not further prepared	134 205	286 661*
200390	Mushrooms n.e.s., preserved, not pickled	n.a.	82 848
080290	Nuts edible, fresh or dried, n.e.s.	222 915	403 243
090610	Cinnamon and cinnamon-tree flowers, whole	95 626	81 332
090620	Cinnamon and cinnamon-tree flowers, crushed or ground	8 531	18 606
110620	Flour or meal of sago, starchy roots or tubers	18 063	10 060
120799	Oil seeds and oleaginous fruits, n.e.s.	62 297	161 428
130232	Mucilages and thickeners, from locust bean, guar seeds	141 335	254 683
130239	Mucilages and thickeners n.e.s.	138 579	374 674
140190	Vegetable materials n.e.s., used primarily for plaiting	39 670	38 181
140200	Vegetable materials for stuffing/padding	n.a.	3 751
140300	Vegetable materials for brush/broom making	n.a.	23 519
140410	Raw vegetable material primarily for dyeing and tanning	31 063	33 855
140490	Vegetable products n.e.s.	63 859	127 767
320190	Tanning extracts of vegetable origin	20 515	50 450
320300	Colouring matter of vegetable or animal origin	152 082	384 133
330129	Essential oils, n.e.s.	312 524	533 464
330130	Resinoids	61 359	37 282
380510	Gum, wood or sulphate turpentine oils	31 232	35 418
380610	Rosin and resin acids	166 133	224 360
410320	Reptile skins, raw	11 252	78 366
430180	Raw fur skins of other animals, whole	44 025	88 240
460110	Plaits and products of plaiting materials	17 198	38 927
460120	Mats, matting and screens, vegetable plaiting material	215 957	196 784
460191	Plaited vegetable material articles not mats or screen	44 732	120 719*
460210	Basketwork, wickerwork products of vegetable material	789 991	968 044
660200	Walking sticks, seat-sticks, whips, etc.	10 769	44 369

* 2001 value (as no longer in HS 2002).
Notes: n.a.: not applicable as this code did not exist in the HS 1992 version.
   n.e.s.: not elsewhere specified.
Source: UN, 2004.

sago flour and wickerwork. These originate in developing countries and were traded as raw materials. Commodities that saw their import values sharply increase are mosses/lichens and foliage for flower bouquets, truffles, other mushrooms, maple syrup, cork, mucilages and thickeners from locust bean (carob), essential oils not elsewhere specified, live animals other than farm animals, natural honey and raw reptile skins. They represent semi-processed products and are mainly produced and traded by developed countries (Europe, North America) and China.

Interpretation of trends
Trends in trade of NWFPs over the past decade need to be interpreted with caution, especially when these items enter the market as ingredients of composite products – a fact that makes their identification difficult in country statistics, complicates the aggregation of their trade value and possibly underestimates their importance. From 1992 to 2002, the value of global trade in related commodities, in current dollars, increased 1.5 times. Trade statistics suggest a modest increase in the total value of NWFPs compared with the growth in overall trade.

Developing countries exported NWFPs as raw materials in the past, but are now processing many of them prior to export. Today the value of trade in NWFPs is mainly from processed products being traded among developed countries (mostly in Europe and North America) and China.

Many NWFPs whose international trade is increasing originate from more intensive production systems, be they in- or outside the forest. More and more, the resources from which these products are derived are being domesticated and cultivated on farms, including important medicinal plants that, until the late 1980s, were obtained exclusively from the wild. In addition to clearing forests for agriculture or to make way for plantations of oil- and cocoa palm and rubber, forests are sometimes cleared to grow bamboo for shoot production in China, cardamom in northern India and *Ilex paraguariensis* for yerba maté in

> **Difficulties associated with collecting, compiling and analysing trade data on NWFPs**
>
> - The term is not included in international commodity descriptions or in product classification systems.
> - Listings that describe or categorize NWFPs within commodities vary considerably, as does their aggregated value, since there is no agreement among countries, agencies or authors on terminology.
> - International commodity nomenclature and product classification schemes are silent on whether products originate from farms or in the forest.
> - Several NWFPs are traded as processed or semi-processed products or as ingredients in other commodities and cannot easily be identified.
> - Changes in product nomenclature in international statistical systems – with codes deleted, merged, split or added – make comparisons difficult over time.
> - Not all countries report accurately on their trade.

Argentina, for example. The potential negative impacts on forest biological diversity of further promoting or increasing trade in NWFPs also need to be clarified.

Prior to promoting commercialization of NWFPs in poverty alleviation programmes, a number of issues need to be carefully considered, including benefit sharing. Experience has shown that increasing trade of these products will not necessarily help poor people, given that the required skills and investment capital are often not available to them (Belcher, 2003). Many NWFPs are competitive only because those who gather the product in forests are paid low wages and often have no other option for cash income. When rural livelihoods improve through farming and industry jobs, rural people no longer want to

collect NWFPs, as happened with cork in Italy and southern France, pine resin in former East Germany and rattan in Malaysia.

Assessing trade in NWFPs is a complex task, mainly because few products appear in classification and nomenclature systems. As recommended by an expert consultation convened by FAO and INBAR, 13 codes for bamboo and rattan products will be added to the HS for 2007 (FAO, 2003d). Similar efforts are required to give prominence to the most valuable NWFPs within other commodities. Such products include nuts, essential oils, mushrooms, oilseeds, medicinal plants, mucilages, colouring matter of vegetable origin, fruits not elsewhere specified and foliage for flower bouquets.

Local uses of NWFPs and their trade within countries have more impact on poverty alleviation and sustainable forest management than international trade. However, the effects of global trade need to be further investigated, given that the most commercially successful products are processed in developed countries and originate from intensive production systems, often outside the forest. ◆

REFERENCES

Arancon, R.N. 1997. *Asia Pacific Forestry Sector Outlook Study: focus on coconut wood.* Asia-Pacific Forestry Sector Outlook Study Working Paper No. APFSOS/WP/23. Rome, FAO.

Ariyadasa, K.P. 2002. *Assessment of tree resources in the home gardens of Sri Lanka.* Bangkok, EC-FAO Partnership Programme on Information and Analysis for Sustainable Forest Management.

Balsiger, J., Bahdon, J. & Whiteman, A. 2000. *The utilization, processing and demand for rubberwood as a source of wood supply.* Asia-Pacific Forestry Sector Outlook Study Working Paper No. APFSOS/WP/50. Rome, FAO.

Bandaratillake, H.M. 2001. The efficacy of removing natural forests from timber production: Sri Lanka. *In* P.B. Durst, T.R. Waggener, T. Enters & T.L. Cheng,

eds. *Forests out of bounds*, pp. 137–166. RAP (Regional Office for Asia and the Pacific) Publication 2001/08. Bangkok, FAO.

Belcher, B. 2003. *NTFP commercialization: a reality check.* Presented at the side event "Strengthening global partnerships to advance sustainable development of non-wood forest products", XII World Forestry Congress, Québec City, Canada, 20 September 2003 (available at www.sfp.forprod.vt.edu/discussion).

Bystriakova, N., Kapos, V. & Lysenko, I. 2004. *Bamboo biodiversity – Africa, Madagascar and the Americas.* UNEP-WCMC Biodiversity Series 19. Cambridge, UK, United Nations Environment Programme World Conservation Monitoring Centre/International Network for Bamboo and Rattan (UNEP-WCMC/INBAR) (available at www.unep-wcmc.org/resources/publications/UNEP_WCMC_bio_series/19.htm).

Bystriakova, N., Kapos, V., Stapleton, C. & Lysenko, I. 2003. *Bamboo biodiversity – information for planning conservation and management in the Asia-Pacific region.* UNEP-WCMC Biodiversity Series 14. Cambridge, UK, UNEP-WCMC/INBAR (available at www.unep-wcmc.org/resources/publications/UNEP_WCMC_bio_series/14.htm).

FAO. 2001. *Global Forest Resources Assessment 2000 – Main report.* FAO Forestry Paper No. 140. Rome (available at www.fao.org/forestry/site/fra2000report/en).

FAO. 2002. *Forests and forestry in Small Island Developing States,* by M.L. Wilkie, C.M. Eckelmann, M. Laverdière & A. Mathias. Forest Management Working Paper No. FM 22. Rome.

FAO. 2003a. *Sustainable forest management and the ecosystem approach: two concepts, one goal,* by M.L. Wilkie, P. Holmgren & F. Castañeda. Forest Management Working Paper FM 25. Rome (available at www.fao.org/forestry/site/20707/en).

FAO. 2003b. *International Conference on the Contribution of Criteria and Indicators for Sustainable Forest Management: the Way Forward (CICI-2003). Report.* Guatemala City, 3–7 February 2003. Rome.

FAO. 2003c. *Workshop on Tropical Secondary Forest Management in Africa: reality and perspectives. Proceedings.* Nairobi, 9–13 December 2002. Rome.

FAO. 2003d. *Proceedings on an FAO-INBAR Expert Consultation on developing an action programme towards improved bamboo and rattan trade statistics,* 5–6 December 2002. Rome.

FAO. 2004. *FAO Forest Products Yearbook 2002.* Rome. (data available at: apps.fao.org/faostat/collections?version=ext&hasbulk=0&subset=forestry).

Ganapathy, P.M. 2000. *Sources of non-wood fibre for paper, board and panels production – status, trends and prospects for India.* Asia-Pacific Forestry Sector Outlook Study Working Paper No. APFSOS/WP/10. Rome, FAO.

IPCC (Intergovernmental Panel on Climate Change). 2004. *Good Practice Guidance for Land Use, Land-Use Change and Forestry.* Geneva, Switzerland (available at www.ipcc-nggip.iges.or.jp/public/gpglulucf/gpglulucf.htm).

ITTO (International Tropical Timber Organization). 2002. *ITTO guidelines for the restoration, management and rehabilitation of degraded and secondary forests.* ITTO Policy Development Series No. 13. Yokohama, Japan.

Jiang, Z. 2002. *Bamboo and rattan in the world.* Shenyang, China, Liaoning Science and Technology Publishing House.

Killmann, W. 1984. Situation of chipboard industry in Pakistan. *Pakistan Journal of Forestry,* 34(2): 65–73.

Killmann, W. 2001. *Non-forest tree plantations.* FAO Forest Plantations Thematic Papers: Working Paper FP/6. Rome, FAO.

Killmann, W. & Fink, D. 1996. *Coconut palm stem processing. A technical handbook.* Eschborn, Germany, Protrade, GTZ.

Killmann, W. & Woon, W.C. 1990. *Oilpalm stem utilization: costs of extraction and transportation.* FRIM Report No. 54. Kepong, Malaysia, Forest Research Institute Malaysia.

Kollert, W., Killmann, W. & Sudin, R. 1994. The financial feasibility of producing gypsum-bonded particle boards from oil palm trunk fibres. In *Proceedings, 3rd National Seminar on Utilization of Oilpalms and Other Palms,* Kuala Lumpur, 27–29 September 1994, pp. 117–137.

Razak, A.M.A. 2000. Recent advances in commercialisation of oil palm biomass. *Malaysian Timber,* 6(3): 12–15.

Ruiz-Pérez, M., Fu, M., Yang, X. & Belcher, B. 2001. Bamboo forestry in China: toward environmentally friendly expansion. *Journal of Forestry,* 99(7): 14–20.

UN. 2004. *UN commodity trade statistics database (UN Comtrade)* (available at unstats.un.org/unsd/comtrade).

WCO. 2004. *Harmonized system.* Brussels, World Customs Organization (available at www.wcoomd.org). ◆

Management, conservation and sustainable development of forests

Sustainable forest management can mean different things to different people. Criteria and indicators to monitor, measure and assess forest trends and conditions have significantly improved understanding of the concept, yet putting it into practice remains a challenge. Practitioners and policy-makers within and outside the forest sector have come to realize that managing forests in a sustainable manner involves the participation of a range of partners to balance trade-offs and resolve conflicts. This chapter examines the similarities and differences between sustainable forest management, as outlined by the "Forest Principles" adopted by UNCED, and the ecosystem approach, as defined by the Convention on Biological Diversity (CBD) and as applied to forests; describes successful forest landscape restoration practices; notes expanding opportunities for forest-based ecotourism in developing countries; identifies issues related to biosecurity, with a focus on invasive species; highlights developments in biotechnology in forestry; and examines international aspects of wildfire management, underscoring the importance of collaborative agreements to assist in cases of fire emergencies.

SUSTAINABLE FOREST MANAGEMENT AND THE ECOSYSTEM APPROACH

"Sustainable forest management", "ecologically sustainable forest management", "forest ecosystem management", the "ecosystem approach" to forest management and "systemic forest management" are among the many terms used to describe concepts and practices that incorporate the three pillars of sustainable forest management – economic, environmental and socio-cultural aspects – to varying degrees.

Recent discussions in the international forest dialogue have focused on the extent to which sustainable forest management and the ecosystem approach as applied to forests are similar, where they differ and how they could be integrated. The UNCED Non-Legally Binding Authoritative Statement of Principles for a Global Consensus on the Management, Conservation and Sustainable Development of All Types of Forests (the "Forest Principles") (United Nations, 1992) outlines the former concept, while CBD defines the latter.

The results of these discussions may have implications both for forest managers and for national planning, monitoring, assessment and reporting. Which approach should forest managers apply – and how? Can countries use the same indicators for monitoring and reporting on progress towards sustainable forest management and towards application of the ecosystem approach to forests?

At the national and international level, clarification and potential integration of the two concepts might enable a better coordination and correlation between the expanded Programme of Work on Forest Biological Diversity of CBD and the Proposals for Action of the Ad Hoc Intergovernmental Panel on Forests (IPF) and the Intergovernmental Forum on Forests (IFF) and thus avoid duplication and reduce the reporting burden on countries. It might also help to clarify the linkages and synergies between National Biodiversity Strategy and Action Plans and national forest programmes.

The concept of sustainable forest management
Sound forest management, taking into account social, cultural, economic and environmental values now and for the future, was widely practised in a number of countries before UNCED and the adoption of the Forest Principles. The concept of sustained yield was applied in forestry for more than a century and, together with watershed management, other soil and water conservation measures and forest protection, has helped maintain the vitality and productivity of production forests. The system of protected areas, which has led to the establishment of a large network of conserved forest ecosystems that now amounts to about 12 percent of the total forest area in the world (FAO, 2001), also long preceded UNCED. International agreement on the Forest Principles provided a basis on which to build a common understanding of sustainable forest management and measure progress.

The Forest Principles state that "forest resources and forest lands should be sustainably managed to meet the social, economic, ecological, cultural and spiritual needs of present and future generations", that "these needs are for forest products and services, such as wood and wood products, water, food, fodder, medicine, fuel, shelter, employment, recreation, habitats for wildlife, landscape diversity, carbon sinks and reservoirs, and for other forest products" and that "appropriate measures should be taken to protect forests against harmful effects of pollution, including airborne pollution, fires, pests and diseases, in order to maintain their full multiple value".

The concept of sustainable forest management has influenced many new initiatives, prompted revisions to forest policies and practices and been widely accepted around the world by forestry organizations at all levels. It continues to evolve through implementation of criteria and indicators processes at the national, regional and ecoregional levels. Extensive collaboration among these processes has resulted in the identification of seven common thematic elements covering the main aspects of sustainable forest management (see Box on page 3). Through the development

and application of indicators for monitoring change, the concept is made operational at the national, as well as local, level.

The ecosystem approach and its application to forests
A key outcome of UNCED was CBD, which has three main goals: the conservation of biological diversity, the sustainable use of its components and the fair and equitable sharing of the benefits from the use of genetic resources. At the second meeting of the Conference of the Parties to CBD (COP-2), delegates agreed that "...the ecosystem approach should be the primary framework of action to be taken under the Convention" (CBD, 1995). The ecosystem approach is based on 12 principles as contained in Decision V/6 of COP-5 to CBD (CBD, 2000).

CBD describes the ecosystem approach as "... a strategy for the integrated management of land, water and living resources that promotes conservation and sustainable use in an equitable way" (CBD, 2000). It also notes that "an ecosystem approach is based on the application of appropriate scientific methodologies focused on levels of biological organization, which encompass the essential structure, processes, functions and interactions among organisms and their environment. It recognizes that humans, with their cultural diversity, are an integral component of many ecosystems".

The term ecosystem "can refer to any functioning unit at any scale. Indeed, the scale of analysis and action should be determined by the problem being addressed. It could, for example, be a grain of soil, a pond, a forest, a biome or the entire biosphere". The concept builds on similar approaches such as the one applied to the management of natural resources by the Man and Biosphere (MAB) Programme of the United Nations Educational, Scientific and Cultural Organization (UNESCO) in the 1970s; the ecosystem management approach, developed in the United States forest sector in the 1980s, and comparable developments in Canada and other countries; and work by the IUCN Commission on Ecosystem Management, the World Wide Fund for Nature (WWF) and other

environmental non-governmental organizations (NGOs).

Comparison of the concepts

A comparison of the two concepts and their underlying principles reveals few differences other than that sustainable forest management deals largely with only one kind of ecosystem – forests – whereas the ecosystem approach addresses a range of ecosystems (Wilkie, Holmgren and Castañeda, 2003). The main points arising from the comparison are as follows.

- Management, conservation and sustainable use of renewable natural resources are the stated goals of both concepts, providing a good example of how two independent processes have developed what is essentially the same vision.
- Both concepts are guided by a set of principles that, although similar, differ slightly in scope. The ecosystem approach principles are, for example, less concerned with the enabling conditions and prerequisites at the national and international levels than the Forest Principles. Some aspects included in the Forest Principles are, understandably, specific to forests and not applicable to other ecosystems and sectors.
- Principles and concepts common to both sustainable forest management and the ecosystem approach include national sovereignty over resources; duty of care (the responsibility for taking care of the environment and preventing adverse environmental impact, even across borders); the "polluter pays" principle; participation; intergenerational equity; conservation of ecosystem structure and functioning; multiple and sustainable use of resources; the need for environmental impact assessments; and equitable benefit sharing.
- While the ecosystem approach appears to deal primarily with ecological and environmental aspects – one of the three pillars of sustainable forest management – the preamble and the rationale of the principles make it clear that social and

economic dimensions are equally important. Recent discussions on sustainable use and benefit sharing within CBD confirm this thinking.

- The few conceptual variations between the two sets of principles stem from different starting points (production forests and forest management versus conservation ecology) but are minimal for all practical purposes. In terms of field application, the differences are likely to be overshadowed by divergent interpretations, local conditions and capacity for implementation.
- As the concept of sustainable forest management has evolved, emphasis has been on what precisely needs to be achieved (specified by criteria) and how outcomes can be measured, monitored and demonstrated (through the monitoring of indicators). The ecosystem approach, a more recent development, has been focusing on the content of the principles, although efforts are under way to provide additional, practical guidance for its implementation (CBD, 2003).

COP-7 to CBD noted that sustainable forest management, as contained in the Forest Principles, can be considered as a means of applying the ecosystem approach to forests. Tools developed in the context of sustainable forest management, including criteria and indicators, national forest programmes, model forests and certification schemes, could potentially help to implement the ecosystem approach. COP-7 also concluded that substantial opportunity existed for those implementing the two concepts to learn from each other (CBD, 2004).

In addition, COP-7 requested that the Executive Secretary of CBD collaborate with the Coordinator and Head of the United Nations Forum on Forests (UNFF) Secretariat and members of the Collaborative Partnership on Forests (CPF) to integrate the two concepts further. In this regard, the ecosystem approach could consider the lessons learned from sustainable forest management, particularly in the application of criteria and indicators. On the other hand, in sustainable forest management greater emphasis could be placed on collaboration within and among sectors; the interactions between forests and other biome/habitat types within a landscape; and biodiversity conservation, in particular through continued development of criteria and indicators and certification schemes.

Integrating the concept of sustainable forest management and the ecosystem approach should lead to using the same indicators to monitor and report progress, thereby reducing the reporting burden on countries. It is also expected to lead to synergies in policy and planning processes at the international and national levels.

Enhanced sharing of information and experiences among practitioners, countries, CBD, UNFF and other members of CPF can improve forest practices at the field level. Moreover, many tools to apply sustainable forest management may be useful in other ecosystems, and focusing on conservation of biological diversity and intersectoral collaboration within the ecosystem approach can help to refine sustainable forest management. Sustainable forest management, particularly in developing countries, can generate socio-economic and financial returns, reduce poverty, increase food security and bring about social equity and sustainable livelihoods. Thus, it represents a viable option among competing land uses.

Planted forests and trees are used to restore an erosion-prone landscape heavily degraded by unsustainable farming practices in New Zealand

D. RHODES

Rather than continue the debate on differences and similarities between the two concepts, efforts should now focus on their implementation, building upon best practices and tools and monitoring progress on the ground to improve national, regional, ecoregional and international policy processes.

FOREST LANDSCAPE RESTORATION

Conventional approaches to the sustainable management and conservation of forests seek to minimize loss of both the extent and quality of the resource. To this end, many national strategies have established networks of forest protected areas and introduced best practices in management of production forests. Many would argue that securing existing forest resources has taken priority over restoring degraded lands at the landscape or holding level where forests and trees have already been lost. Consequently, until recently, restoring forest resources has focused on establishing planted forests for the production of industrial roundwood, revegetating heavily impacted sites such as mines, quarries and landfills and restoring ecologies to enhance connectivity between sites of high biodiversity importance.

Notwithstanding the importance of ensuring that countries protect and manage their forests in a sustainable manner, there is a growing realization that such a strategy alone may be insufficient to guarantee a healthy, productive and biologically rich forest estate for the longer term. In some regions, so much forest has been lost or degraded that the supply of goods and services on which local, and sometimes national, economies depend is threatened. It is also now well established that fragmentation can exacerbate the vulnerability of many forest types to threats such as wildfires and invasive species. Finally, the impact that climate change will have on both intact and fragmented forest stands presents a serious challenge to optimizing the resilience and resistance of forest resources to global warming.

Forest landscape restoration aims to regain ecological integrity and enhance human well-being in deforested or degraded forest

landscapes (Maginnis and Jackson, 2002). The process brings stakeholders together from different sectors to put in place a variety of land-use practices that will help to restore the social, environmental and economic functions of forests and trees across the landscape. Since the launch of the Global Partnership on Forest Landscape Restoration (see Box below) at the sixteenth session of the FAO Committee on Forestry (COFO) in March 2003, organizations and governments have been exploring the concept as a possible complement to the management and protection of forest resources. Although it is not a new idea, its novelty lies in addressing and balancing trade-offs at the landscape level and its pragmatic rejection of the need to return modified forest landscapes to their original pristine state. Forest landscape restoration is carried out under the assumption that improving the flow of forest goods and services requires a balance between livelihoods and nature protection, and that this is best achieved within dynamic, multifunctional landscapes.

Global Partnership on Forest Landscape Restoration

The Global Partnership on Forest Landscape Restoration is a growing network of governments, international and non-governmental organizations and communities that are working to raise the profile of forest landscape restoration as a model of how the international forest community can link policy with practice. Under the partnership, more than a dozen national and regional workshops have been held, and several others are planned, to share experiences and to develop and implement practical next steps. An international expert meeting on forest landscape restoration will be convened in 2005 to review lessons learned and plan further coordinated action.

Further information on the global partnership is available at www.unep-wcmc.org/forest/restoration/globalpartnership.

Since a key objective of landscape restoration is to get the right blend of approaches at the right scale to enhance the supply of forest goods and services, efforts are not limited to, nor do they exclude, particular site-based technical interventions. Forest landscape restoration consists of a flexible package of these interventions, which include ecological restoration, natural forest management, regeneration of secondary forests, afforestation and reforestion of planted forests, woodland and rangeland management and planting of trees outside forests, including agroforestry and urban and peri-urban forests. The combined result should be a landscape mosaic of forests and trees that contribute to livelihoods and to sustainable land use and development.

Forest landscape restoration in practice

A restored landscape might consist of areas that are protected for watershed management and nature conservation, linked by regenerated native forests along rivers and streams. The landscape may also include well-managed natural or planted forests for production of wood and non-wood forest products for industrial purposes. Complementing these can be a diverse range of agroforestry plantings and trees outside forests, which provide valuable goods and services to smallholders. The restored landscape could also offer opportunities for recreation, amenity and tourism. Enhancing the multidisciplinary and intersectoral nature of landscape restoration depends as much on the needs and aspirations of local stakeholders, on institutional and land-tenure arrangements and on the prevailing land-use policy framework as it does on biotic factors such as residual soil fertility and remnant forest species diversity, abundance and distribution.

Support from local stakeholders is fundamental to the success of any restoration activity. Public and private investors, as well as smallholder landowners, need to be allowed to invest in natural and planted forests and trees and be confident that they will receive benefits. When restoration helps meet the broader demands of society by providing such services as biodiversity conservation, carbon sequestration and watershed protection, incentives or new market mechanisms may be required to compensate local people. In addition, governance issues must be addressed, including the need for consistent and enabling policy and for legal and regulatory frameworks that contain clear commitments on land rights and on forest ownership and use. Traditional practices and institutions can also have a significant role.

In restoring forest landscapes it must also be recognized that objectives may shift over time. While long-term aims may be to increase the resilience, diversity and productivity of land-use practices and to conserve biodiversity, short-term interventions may be required to meet immediate needs, for example for production-based benefits. The following examples illustrate the results of restoration initiatives.

In the United Republic of Tanzania, the Sukuma people in Shinyanga had a strong pastoralist tradition and relied on *Acacia* woodland enclosures, or *ngitili*, to provide dry-season fodder and a range of other essential goods and services. However, as a result of tsetse fly eradication schemes, the conversion of land for cash crops and state-sponsored collective farming, by 1985 only about 1 000 ha of *ngitili* remained, and land degradation had become a serious issue. A government-sponsored soil conservation project set out to work with traditional land-use systems and to build on institutional structures. These efforts coincided with a relaxation in the rules governing collective farming. By 2000, the area of *ngitili* had increased to more than 250 000 ha. Although restored patches range between 10 and 200 ha, their cumulative effect has dramatically transformed the Shinyanga landscape (Barrow *et al.*, 2002). The recovery of landscape-level forest functionality, in this instance, had little to do with formal planning processes or tree planting. Rather, it was the lifting of land-use constraints and empowerment of local traditional institutions that allowed *ngitili* to flourish.

In 1970, the 50 000 ha of even-aged Sitka spruce (*Picea sitchensis*) plantations in Kielder Forest

supplied 5 percent of the United Kingdom's softwood requirements (Global Partnership on Forest Landscape Restoration, 2004). Although the forest was successful in terms of timber production, the lack of public access and the perceived deterioration of the environmental and wildlife habitat value of this publicly owned estate were an increasing cause of dissatisfaction. The Forestry Commission modified the forest while maintaining its productive capacity, increasing the proportion of native broad-leaved species to 8 percent, up from 1 percent in 1980, ostensibly for aesthetic and habitat purposes. Moreover, it altered restocking practices in 20 percent of the harvested compartments to improve biodiversity conservation. In this way, the Forestry Commission enhanced social and environmental attributes at the landscape level while delivering 1 400 tonnes of roundwood per day on a sustainable basis. Greater efforts were made to include people in the restructuring and management of Kielder Forest so that, although the workforce fell from 2 000 to 260 employees over 50 years, the number of visitors rose to half a million per year, revitalizing the local economy through tourism and related services.

From the mid-1970s onwards, a combination of poor harvesting methods, shifting agriculture and fire degraded large tracts of forests in Asia and the Pacific to the point where there was little potential for wood crops or ecological services such as carbon sequestration, watershed protection and biodiversity conservation. Without remedial action, these degraded forest lands were likely to be converted to other uses. In the late 1990s, the Forestry Research Support Programme for Asia and the Pacific (FORSPA) established a forest rehabilitation network, which launched pilot sites in Cambodia, the Lao People's Democratic Republic, Papua New Guinea, Sri Lanka and Viet Nam. Management protocols have been developed between forestry specialists and local communities, taking into account the unique social, environmental and economic conditions and incorporating scientific and traditional knowledge in restoration initiatives. The network is raising interest in forest landscape restoration in the region and

Searching for excellence in forest management

The Asia-Pacific Forestry Commission (APFC) has recently finalized an initiative entitled "In search of excellence: exemplary forest management", launched in November 2001. Individuals from across Asia and the Pacific were invited to nominate forests that they perceived to be well managed and to elaborate on management aspects they considered exemplary.

"In search of excellence" identifies:

- examples of good forest management across a broad range of forest ecotypes in the region, covering a variety of objectives, ownership structures and sizes of forest area;
- practices that show promise for the future and for other areas;
- perceptions of what constitutes good forest management.

The invitation to nominate forests was extended via Web sites, newsletters and brochures over several months. Workshops were also convened in nine countries, providing opportunities for participants to debate the elements of good forest management.

More than 170 nominations of both planted and natural forests were received from 20 countries. They ranged in area from less than 20 to nearly 2.5 million hectares and included forests managed for watershed protection, biodiversity conservation, timber or non-wood forest products, recreation, agroforesty, tourism and rehabilitation. Submissions covered state-owned, private and community forests as well as joint ventures.

Ten technical experts selected 30 forests for in-depth case studies. These were examples that stood out for specific management aspects and demonstrated innovation in the face of challenge. Emphasis was placed on identifying a variety of management experiences that characterized a range of objectives across several countries.

The case studies were published in April 2004, together with an analysis of commonalities and differences in management among nominated forests. The findings revealed that there is no "right" way to manage forests and that approaches vary according to cultures, local conditions and management objectives. However, common elements were evident in the management of most forests in the sample.

Excellence tended to be defined in terms of outstanding forest practices based on scientific principles as well as on participatory and transparent management. It was often identified in the context of impressive biophysical changes (e.g. rehabilitation of degraded areas, reduced soil erosion, enhanced water quality and yield) or positive socio-economic changes (e.g. increased incomes for local people, improved availability of forest products, enhanced understanding and appreciation of forest health protection). The involvement of stakeholders in decision-making and the management of forests for multiple benefits were also recognized as important elements.

FAO recently carried out a similar exercise in Central Africa, in collaboration with the Interafrican Forest Industries Association (IFIA), the World Wide Fund for Nature (WWF), the African Timber Organization (ATO), the International Model Forest Network Secretariat (IMFNS), the International Tropical Timber Organization (ITTO) and the World Conservation Union (IUCN). Case studies of this initiative were published in FAO Forestry Paper No. 143 (FAO, 2003a).

is facilitating the exchange of information, experiences, technology and expertise.

Ideas for the way forward

Case studies and regional workshops evaluating the role of forests and trees in urban and rural landscapes consistently and clearly point to the need for:

- decentralized, participatory and multidisciplinary approaches to policy, planning, management and monitoring;
- maintenance of forests and trees as integral components of the landscape;
- supportive institutional frameworks and greater intersectoral collaboration;
- integrated approaches to balance short-term needs for food and livelihoods with long-term needs for environmental services, including biodiversity conservation;
- dissemination of knowledge and technology concerning the role of forests and trees in restoring wider landscapes, through national and international networks;
- sound extension and technical support systems and demonstrations of forest landscape restoration at work;
- interventions that reflect the unique physical, cultural, social, political, environmental, economic and institutional conditions of each landscape.

FORESTRY AND ECOTOURISM: EXPANDING OPPORTUNITIES IN DEVELOPING COUNTRIES

The recent boom in nature tourism and ecotourism provides emerging challenges and opportunities for forest management. As the world's largest employer, the tourism industry directly or indirectly generates more than 200 million jobs, or 8.1 percent of total employment, globally. The value of travel and tourism exceeds US$4.2 trillion annually, or more than 10 percent of global gross domestic product (GDP) (WTTC, 2004).

Nature tourism, of which ecotourism is a segment, accounts for a sizeable, if somewhat uncertain, share of this large industry. While there is no accepted definition of "nature

tourism", it is generally considered to encompass activities that relate to or depend on natural attractions, including outdoor sports, hunting, fishing, canoeing and backpacking. The International Ecotourism Society defines ecotourism as "responsible travel to natural areas that conserves the environment and sustains the well-being of local people". Although widely accepted, the definition is not functional for gathering statistics, making it impossible to determine accurately ecotourism's share of total tourism (measured in terms of tourists, expenditures, employment or contribution to GDP). However, ecotourism is already a profitable business, and most analysts agree that it is the fastest-growing segment of the industry.

Some claim that as many as 40 to 60 percent of all international tourists are nature tourists (Fillion, Foley and Jacquemot, 1992), while most estimate the percentage of ecotourists at between 10 and 20 percent because the term is defined more narrowly (Pleumarom, 1994; Ananthaswamy, 2004). Much of nature tourism and ecotourism focuses on forests. From bird-watching to canopy walks, forest treks and wildlife viewing, growth in the sector means it will increasingly influence how forests are used around the world.

Environmental advocates and development experts are tapping the huge potential of tourism to advance both conservation and rural development, especially in areas where logging is restricted or undesirable. Many people perceive ecotourism as capable of delivering substantial environmental, social, cultural and economic benefits at the local and national levels. It provides a means by which people can use forests and wildlife to generate income without extracting resources or degrading the environment, and it presents a strong incentive to protect the resource. If managed properly, ecotourism can generate income and employment for rural communities faced with few alternative livelihood opportunities. Examples of ecotourism and its potential can be found in every region of the world (see Box on page 28).

Potential of ecotourism: some examples

- The Kenya Wildlife Service estimates that 80 percent of Kenya's tourists are drawn by the country's wildlife and that the tourism industry generates one-third of the country's foreign exchange earnings (Kenya Wildlife Service, 1995).
- Domestic and international travellers make more than 275 million visits a year to the 388 recreation areas administered by the United States National Park Service (United States Department of the Interior, 2004), generating direct and indirect economic benefits for local communities of more than US$14 billion annually and supporting almost 300 000 tourist-related jobs (Tourism Works for America Council, 1997).
- Prior to the civil war in Rwanda, tourists visiting the country's mountain gorillas provided more than US$1 million in annual revenues, enabling the government to fund anti-poaching patrols and employ local residents (Gossling, 1999). Tourism is once again on the upswing, with hundreds of foreign visitors a month handing over US$250 each to see the gorillas.
- More than half of all international visitors to Nepal include a trip to at least one national park. Before civil strife reduced numbers, more than 80 000 tourists visited Royal Chitwan National Park, and 50 000 trekkers visited Annapurna Conservation Area each year (Yonzon, 1997).
- The more than 60 000 visitors a year to the Galapagos Islands contribute in excess of US$100 million to Ecuador's economy (Charles Darwin Research Station, 2001).

Ecotourists are seeking nature in its pristine state, which is often found only in remote destinations. Under the UNESCO Nam Ha Ecotourism Project in the Lao People's Democratic Republic, trekkers are brought to specific tribal villages, which receive US$0.50 per tourist and then use the money to buy medicine, pay for schooling and improve community welfare. Locally recruited guides earn US$5 per day – an exceptional wage by Lao standards – leading tourists and sharing their insight and knowledge of the forest and its wildlife. Revenues have reduced illegal logging and hunting and have improved health conditions for local people (Gray, 2004).

Often, however, mass tourism in natural areas can have a devastating impact. Without measures to ensure otherwise, activities destroy the environment, disrupt social structures and leave few economic benefits for local people. Recent studies indicate that some ecotourism previously thought to be benign stresses wildlife, disrupts breeding patterns and changes the behaviour of wild animals (Ananthaswamy, 2004).

Environmental considerations

Care must be taken to ensure that the very features that provide the basis for attracting tourists are not damaged or destroyed in accommodating visitors' physical needs and comforts. Disturbances to the local ecology – rubbish and waste disposal, cutting of trees for fuelwood, lodges, access routes and communication facilities, for example – are obvious at many sites, including major parks and protected areas.

The overuse of popular ecotourism sites often results in erosion of trails and riverbanks, water pollution, destruction of vegetation and loss of species. Problems can usually be attributed to lack of planning, failure to develop and implement management plans, inadequate monitoring and control mechanisms, low participation of residents living in or near sites and divergent priorities of government agencies, the tourism industry and local populations.

*Growth in the ecotourism
sector will increasingly
influence how forests are
used around the world*

As part of sound management planning, all potential effects of tourism on the ecosystem should be assessed, not only those likely to affect the species that attract visitors to the site. Solid baseline data are also essential to monitor changes that may occur as the industry develops. The concept of carrying capacity, with its physical, social and ecological components, should be taken into account as well.

Significant progress has been made in recent years in the design, construction and management of eco-friendly tourist lodges. Such facilities emphasize the use of local construction materials, careful waste and rubbish disposal, water conservation, and solar power and water heaters. Guides also play a crucial role in protecting the environment by ensuring that tourists do not encroach on sensitive areas, collect threatened or endangered plants or disturb wildlife. Successful ecotourism therefore requires the recruitment and training of guides who manage and influence the behaviour of tourists.

Socio-cultural considerations

Tourism of all types, including ecotourism, offers opportunities for people of diverse backgrounds and cultures to exchange views, develop friendships and gain a better understanding of others. On the other hand, tourism can highlight differences, fuel animosities and lead to clashes of cultures, especially when wealthy tourists visit isolated or less-developed regions.

Ecotourism, if not controlled, can rapidly stretch the social fabric of remote forest villages and cultures. Inflation of local prices, loss of ancestral lands, behavioural and value changes, prostitution, drug abuse and diseases

are real threats. Many indigenous people in developing countries have only recently begun to experience the impact of a market economy. As some members of the community are quicker than others to earn money from tourism, they may rival traditional leaders and elders in terms of prestige. Their income may be many times higher than what a villager can earn by conventional means, perhaps leading to jealousy and violence. Other negative results include begging and hostility towards tourists.

Economic considerations

The extent to which ecotourism is able to conserve forests and develop rural areas is largely contingent on capturing revenues to manage parks and other forest land and discourage destructive practices. Too often, however, the money that ecotourism generates goes to other countries, providing little incentive to protect the resource. This type of leakage occurs in the form of payments to tour operators, airfare, foreign-owned accommodation and non-local supplies and food. The World Bank estimates that only 45 percent of tourism's revenue worldwide stays in the host country, and a study of the popular Annapurna region of Nepal found that only 10 percent of tourism expenditures benefited the local economy (Martinoli and Fiore, 1999).

Increasingly, governments are demanding that parks and protected areas generate sufficient revenue to cover the cost of their management through such means as entrance and user fees and concession licences. Thailand, for example, expanded its infrastructure, upgraded facilities, intensified marketing efforts and increased entrance fees. Such approaches are not without controversy, however. Park managers, trained in resource protection, are often apprehensive about problems that rising numbers of tourists bring. Tourists, on the other hand, sometimes resent paying high fees, especially under

Assisted natural regeneration: a simple technique for forest restoration

The term "assisted natural regeneration" was first coined in the Philippines, where the approach has been used to restore forest cover to *Imperata cylindrica* grasslands by working with and building on the principles of natural plant succession.

Known locally as *cogon* in the Philippines and *alang-alang* in Indonesia, *Imperata cylindrica* is an aggressive grass that covers more than 50 million hectares of land in Asia and the Pacific – land that was mostly covered by forests originally (Garrity *et al.*, 1997). *Imperata cylindrica* is highly flammable, and frequent fires prevent further succession and the natural return of forest cover. However, if *Imperata* grassland does not burn, it will naturally and gradually return to forest as pioneer trees and shrubs eventually grow above *I. cylindrica* and outcompete it for light and water.

There are a variety of techniques for assisted natural regeneration, depending on reforestation objectives, site characteristics and resources available. In general, however, it involves:
- protecting against fire and grazing;
- suppressing *I. cylindrica* and other fire-prone grasses;

- weeding, mulching and applying fertilizer, if needed, to rootstock and young tree seedlings that sprout from seeds carried by natural dispersal agents.

While fire prevention is a key element in rehabilitating *Imperata* grasslands, effective techniques for suppressing this and other weeds have been discovered more recently. In the Philippines, wood planks or bamboo poles are used to press *I. cylindrica* to the ground to slow its growth and reduce its regenerative capacity. Covered grasses in the lower layers quickly die, allowing tree seedlings and saplings to grow and shade the grasses. This simple process also reduces flammability, as air does not circulate well in the compacted grass (Friday, Drilling and Garrity, 1999).

Advantages of assisted natural regeneration relative to conventional reforestation include:
- regeneration of indigenous species;
- restoration of biological diversity and ecological processes;
- lower costs because of the elimination or reduction of seedling production, transport, planting and replanting activities;

two-tiered pricing schemes that charge local residents considerably less.

Expanding tourism in parks and protected areas is doubly taxing if revenues revert to the national treasury. Ecotourism income should help to improve the management of forest areas on which tourism is based but is often not available to the agencies concerned. In Costa Rica, for example, only about one-quarter of the park service's budget comes from fees – not enough to manage and protect its numerous sites. The remainder must come from donors and government allocations.

Challenges
While ecotourism offers good reason to conserve forests and stimulate rural economies, it is not a panacea. Experience has revealed that it succeeds only under certain conditions, and resource managers and development officials would do well to avoid unrealistic expectations.

Ecotourism requires sites that have attractive natural features, such as wildlife in sufficient abundance for tourists to spot, unique plants, waterfalls, mountains and beautiful scenery. Comfortable accommodation, safe hiking trails, good information and visitor-friendly facilities are also important. While a few ecotourists are willing to endure harsh conditions, most prefer a mix of adventure and luxury. Balancing the two requires knowledge of tourist preferences and substantial investment capital.

While ardent ecotourists relish the idea of travelling to remote destinations, most

- easy implementation, requiring no costly tools or skilled labour;
- minimal soil disturbance;
- natural selection and succession of trees appropriate to prevailing conditions.

Experience in Indonesia and the Philippines indicates that successful application of assisted natural regeneration requires the participation of local people and mechanisms to promote the equitable sharing of benefits. While excellent potential exists to apply assisted natural regeneration more widely, constraints include:

- a lack of knowledge of ecosystem dynamics, including the requirements for natural regeneration of species;
- lack of experience in implementing approaches and techniques;
- weak policy and incentive systems with regard to land tenure and equitable distribution of benefits derived from restoring forest diversity (Sajise, 2003).

*In the Philippines, wood planks are used to press **Imperata cylindrica** to the ground to reduce its regenerative capacity, allowing tree seedlings and saplings to grow and shade the grasses*

BAGONG, PAGASA FOUNDATION/E. CADAWENG

do not have the time, desire or money to do so. Sites, therefore, should be accessible – but not too accessible. Moreover, as with other forms of tourism, ecotourism is highly sensitive to perceived risk and physical danger, especially those associated with civil strife, war and terrorism. As experiences in Nepal and Rwanda have shown, thriving businesses can fail rapidly when tourists feel threatened.

Tourism is also influenced by world economics. Middle and upper-middle classes in developed countries are the most sought-after clients because of their purchasing power. In addition, currency exchange rates, political considerations and cultural perceptions influence the decision to travel.

As a highly competitive business, ecotourism requires effective management and marketing skills – skills that are often lacking in rural communities. While projects, donors and NGOs can provide short-term support, local capacity must be built if the industry is to benefit local people. Residents and affected communities also need to be involved in efforts to develop ecotourism, to understand the implications of such development, to benefit from activities and to negotiate with outsiders as equals. Some countries have policies that provide for partial reimbursement to residents for the costs of establishing protected areas.

Ecotourism planners also promote the sale of local handicrafts, the use of local accommodation and training programmes to enable residents to

fill positions as tour guides, lodge managers and park employees (Vanasselt, 2001).

To conserve the natural resources on which ecotourism is based, small-scale tourism is often recommended although it generally brings only small-scale benefits, including seasonal and low-paying jobs. Thus, a major challenge is to identify a scale of ecotourism that will provide profits to local communities without jeopardizing forests and other natural resources.

The recent proliferation of ventures claimed to be ecotourism – many of which harm the environment and fail to provide local benefits – has led to calls for certification of the industry. As with forest certification, the plethora of schemes for certifying ecotourism businesses is resulting in consumer confusion, little recognition of labels and lack of understanding of certification processes. Some businesses claim that certification improves performance, but a difference in the market is not yet apparent. Efforts are ongoing to harmonize processes and raise tourist awareness of certification (Chafe and Honey, 2004).

In conclusion, ecotourism is a highly competitive business that requires considerable capacity to succeed. Most countries are not realizing the full potential of this segment of the industry, are not making effective use of the revenues it generates and are not providing adequate support to develop the sector. With a few exceptions, the forestry profession does not look upon ecotourism as a forest management strategy, so benefits from its successful development tend to accrue to other sectors. Much more could be done to sensitize foresters to the need to include ecotourism within management regimes.

Further information on ecotourism can be found at www.ecotourism.org.

BIOSECURITY AND INVASIVE FOREST TREE SPECIES

Concern over the potential negative impact of the introduction of new species, breeding and the use of genetic modification (GM) has led to increased attention paid to the need to develop regulatory frameworks and policies to manage environmental and biological risks. The management of such risks, generally referred to as biosecurity or bioprotection, is of direct relevance to the sustainability of agriculture, food safety and the health of the environment, including the conservation of biological diversity. In forestry, a recent focus has been on invasive forest tree species (see Cock, 2003; FAO, 2003c).

In addition to possible loss of native species resulting from the spread of introduced tree species, the introduction of new tree genotypes (non-local provenances or genetically improved planting stock) could have an adverse impact as a result of what is sometimes referred to as genetic pollution – the creation of hybrids and the loss of gene pools that may have acquired specific characteristics through local adaptation. To date, however, few studies and few recorded instances of such results are found in the forest sector. Information is also scarce on possible negative effects of the introduction of other species into forest ecosystems, including biological control organisms and mycorrhizae.

Introduced forest tree species can help sustain national and local economies and be of significant value to the environment and to society. However, when insufficient consideration is given prior to use and when on-site management is neglected, some species may invade adjacent areas, giving rise to a number of problems (Robbins, 2002). Moreover, with global trade increasing, greater movement of people and overstretched quarantine services, the number of accidental introductions of potentially invasive forest tree species is expected to rise.

Global information on forest tree and shrub species that have become invasive is inadequate and subject to interpretation because the contexts in which studies have been carried out vary, terminology is unclear and concepts may overlap as in the case of "invasive" (an introduced species that, when unmanaged, invades surrounding habitats) and "naturalized" (an introduced species that has adapted locally, is well established and forms an integral part of the flora of a country or region). Lack of agreement on terms such as "introduced",

FIGURE 2
**Classification of 1 121 tree species
according to geographical distribution
and invasive behaviour**

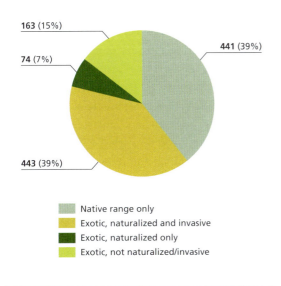

163 (15%)

74 (7%)

441 (39%)

443 (39%)

☐ Native range only
☐ Exotic, naturalized and invasive
☐ Exotic, naturalized only
☐ Exotic, not naturalized/invasive

Source: Haysom and Murphy, 2003.

"alien" and "exotic" and the subjective values attached to them add to the confusion and increase difficulties in assessing the extent and impact of the undesirable spread of forest trees.

Of the more than 1 100 tree species included in a recent survey (Haysom and Murphy, 2003), those outside their native range were classified according to their reported degree of invasiveness (Figure 2). Among those categorized as invasive were 282 species used in forestry. A further 40 were reported as naturalized but not invasive. Both angiosperm and gymnosperm invasive species were identified. In decreasing order, most invasive forest tree species occurred in the families Leguminosae, Pinaceae, Myrtaceae, Rosaceae and Salicaceae.

According to the study, invasive tree species were reported with various intensities in all regions reviewed: Africa, Asia and the Pacific, Australasia, Europe, North America and South America. The largest number occurred in Africa (87 species) and the lowest in Europe (12) and Asia (14). Most species were invasive in only one region, and even those most frequently

considered invasive were not reported to have adverse impacts in all countries where they had been introduced. Most tree species reported to have become invasive in new habitats originated in Asia; the Pacific was the origin of the fewest. However, scant information was available on the history of the introductions or on the subsequent use and management of the trees.

Again according to the study, most invasive tree species were reported in countries and regions where investments to catalogue introductions and conduct research on their impact were high, for example, Canada, Puerto Rico, South Africa and the United States. On the other hand, gaps in information were evident in Africa, Asia and parts of South America.

Risks associated with invasive species, including plants, animals, fish, microbes, pests, insects and diseases, are addressed in the work programme on invasive alien species of CBD and in campaigns and projects of IUCN and other NGOs. However, in some circumstances invasiveness can be a desirable trait, for example, for combating desertification or rehabilitating degraded lands.

FAO convened a Technical Consultation on Biological Risk Management in Bangkok, Thailand, in January 2003, which addressed biosecurity related to food and agriculture (FAO, 2003b). An Asia-Pacific Forest Invasive Species Conference held in Kunming, China, in August 2003 laid the foundations for the establishment of an Asia-Pacific Forest Invasive Species Network under the auspices of the Asia-Pacific Forestry Commission (APFC). The network was formally launched in April 2004. It shares information on forest invasive species and facilitates access to expertise and resources such as education and training facilities and courses.

More decision-makers and professionals should be made aware of the need to evaluate the consequences of introducing new tree and shrub species, especially since tree species that provide useful products and services in one sector can be considered harmful in another. A multisectoral approach is therefore needed to assess impact from different perspectives and to identify management options that balance positive

Biosecurity issues addressed by UNFCCC

During COP-9 to UNFCCC in December 2003, several countries sought to eliminate the use of potentially invasive alien tree species and genetically modified organisms in afforestation and reforestation projects under the Clean Development Mechanism (CDM) of the Kyoto Protocol. The meeting ultimately agreed that the host country of any such project would decide on the use of such trees and the investor country could either accept or reject the resulting carbon credits (UNFCCC, 2003).

and negative aspects. On the positive side, the introduction of faster-growing tree species may enhance carbon sequestration, provide much needed woodfuel and other products, stabilize soils and protect agricultural lands. On the other hand, trees that become invasive create problems in the management of grassland pastures and, in some instances, of agricultural lands and natural or planted forests. Of particular concern is their effect on ecologically fragile natural or semi-natural habitats, such as riparian and wetland systems. A study conducted in the fynbos vegetation region of South Africa (Nyoka, 2003) revealed that introduced invasive trees caused substantial losses in native biodiversity and greatly diminished runoff in water catchments, allegedly affecting South Africa's water supply and requiring expensive controls.

At the local level, the longer the invasiveness of a species remains undetected, the less chance there is for successful intervention. Fewer options will exist for containment through management or for control through eradication, and the costs of interventions will rise over time.

Although it is difficult to predict which species might cause serious damage if introduced, those that are known to have caused problems when brought into other parts of the world provide the best guide for evaluating risks. Thus, access to reliable information and better knowledge of economic and environmental impact are

critically important. Clarification of concepts, terms and definitions at the international level is also a priority, as is reaching agreement on assessment methodologies and on data to be collected at national and local levels to help evaluate and manage risks.

As mandated by member countries, FAO is compiling a glossary of terms and definitions related to biosecurity in food and agriculture, forestry and fisheries. In addition, the Organization is developing databases on introduced, naturalized and invasive species. They can be accessed through the FAO portal on food safety, animal and plant health at www.fao.org/biosecurity.

BIOTECHNOLOGY IN FORESTRY

The term biotechnology refers to developing or using living organisms to produce, alter or improve a product or a living organism for a specified purpose. It encompasses conventional breeding, including plant and animal domestication from prehistoric times, and modern innovations that focus on a portion of a biological system (Yanchuk, 2001). Most public research involving forest biotechnology relates not to GM, but to tools for studying and characterizing the biology and diversity of forest tree species, populations and individuals, or for propagating forest trees. More than two-thirds of non-GM biotechnology research in forestry uses only four genera – *Pinus*, *Eucalyptus*, *Picea* and *Populus*. Research is carried out in all regions of the world, with significant differences among regions and objectives (Figure 3). More than two-thirds of activities on genetic diversity and marker-assisted selection are carried out in Europe and North America, while 38 percent of research programmes using advanced propagation technology are in Asia.

Genetic modification – the transformation of organisms by the insertion of one or more isolated genes – has been subject to passionate debates, most recently over the commercialization of new genotypes (Cock, 2003). Some scientists and some members of the public are worried about risks associated with gene transfer to native populations

FIGURE 3
Forest biotechnology research by region, excluding genetic modification

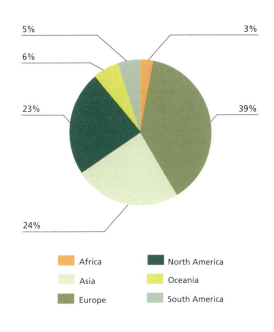

5%

3%

6%

23%

39%

24%

Africa — North America
Asia — Oceania
Europe — South America

Source: FAO, 2004.

(genetic pollution) and about environmental impacts. Although genetic pollution or displacement of native species may also occur with conventionally bred varieties or exotic species, there is uncertainty about the effects of releasing organisms that were obtained by breaking natural barriers that have prevailed in conventional breeding to date. Other concerns include consumer health (although these concerns are less apparent than for agricultural crops) and the equitable sharing of costs and benefits.

While the tools for genetic modification in forestry are mostly the same as those used in the agriculture sector, the potential applications, benefits, impacts and public perceptions differ significantly where forest trees are concerned (El-Lakany, 2004). These differences stem from the social, cultural and environmental aspects of forests. They also stem from the fact that forest trees have only recently been domesticated, in contrast to most agricultural crop species. Many forest trees are still in their wild (unimproved)

state or are removed only one or two generations from their ancestors through breeding programmes.

To improve the amount of reliable information on biotechnology in forestry, FAO is now carrying out the first global review, including developments and applications of genetic modification technology (FAO, 2004). Preliminary findings suggest that, as of 2002, only one country (China) was growing genetically modified forest trees (poplar clones) on an area of less than 500 ha. *Populus* is the genus of forest tree in which genetic modification has been studied most widely, although some research has been reported for 19 genera of woody plants.

Almost half of all research on genetic modification in forest trees takes place in the United States of America, with most of the remainder in other developed countries. However, the technology is growing rapidly, and some of the more advanced developing countries are quick to adopt it.

Most first-generation traits under examination (e.g. pest resistance and herbicide tolerance), with the exception of wood quality traits, derive from research in agricultural crops and are of interest mainly for potential commercial wood production. However, developing, testing and approving genetically modified forest trees for wider use may entail high costs and significant time frames because of the difficulties associated with assessing risks in such long-term crops. Other applications of GM could be found in forest conservation activities, including the recovery of valuable ornamental and urban shade tree species that have succumbed to insects and disease. Another use of GM technology, often overlooked but perhaps the most important, is in basic research on tree biology for better understanding of gene functioning and the characters that genes control.

In many countries, the private sector is undecided and reluctant to communicate its intentions with regard to the deployment of genetically modified trees. While companies may fear that failing to engage in genetically modified organism (GMO) research could mean lost opportunities, they recognize the power of

public opinion and are aware that widespread opposition to genetically modified forest trees poses a commercial risk in a number of countries.

The economic rationale for employing GMOs in forestry has not yet been clearly demonstrated because the monetary value of forest products in global trade is far less than that of agricultural products. Many planted forests grow in countries where improved genetic material and appropriate silvicultural procedures are not used. The success of tree improvement programmes over the past 50 years suggests that there is scope for enhancing productivity and yields on a sustainable basis using conventional forest tree breeding. This is the situation at present, however, and does not imply that the application of GM technology to forest trees will not be advantageous.

As a relatively new tool in forestry, GM technology has potential benefits and drawbacks but is not intrinsically good or bad. Since its usage is technically possible, it should be studied and regulated on a case-by-case basis. Genetic modification in forestry is more than a technical issue. Socio-cultural values and the multiple uses of forests need to be taken into account, and public acceptance is necessary, if genetically modified forest trees are to be deployed.

Keeping an eye on developments

Whether it is government or the private sector that stimulates advances in biotechnology, forest planners need to become more aware of its potential and drawbacks and to consider these aspects when developing future forest management strategies.

The distinctive nature of forest trees and their importance in ecosystems make risk assessment a critical issue in the deployment of many biotechnologies. Thus, national and international agencies need to address the management of such risks from an intersectoral perspective.

Developed and developing countries have different priorities, capacities and applications for biotechnology. However, developing countries could make rapid technological gains and improve their capacity as long as economic

opportunities were available and regulatory frameworks were in place.

Given the high cost of genetic biotechnology and the expected growth in demand for quality industrial wood over the next 30 years, industry will likely focus on intensification and on high-yield plantations. The forest sector needs to monitor developments in GMOs in agriculture because regulations for crops are likely to be adapted for forest trees.

GM and other biotechnologies may have a role to play in plantation forestry in some countries. However, because some 95 percent of the world's forest area is natural or semi-natural, the area planted with genetically modified forest trees is likely to remain small.

FAO intends to continue monitoring biotechnology, including GM, in forestry at the global level as well as to make objective, updated and reliable information available.

WILDLAND FIRES

Much public and media attention is given to uncontrolled fires in forests, other wooded lands and other lands – generally referred to as wildland fires. Since many incidents are not monitored or documented, the absence of reliable assessments of damage and impact hinders decision-making. For this reason the Global Fire Monitoring Center (GFMC) and the Global Observation of Forest Cover Fire Implementation Team have called for international joint efforts to launch an operational space-borne fire monitoring system that will allow real-time and complete coverage of wildland fire events and fire impact around the world (Ahern, Goldammer and Justice, 2001).

The total global area burned in 2002 and 2003 – of which about half was in Africa – appears to be comparable to long-term averages, in the annual range of 300 to 400 million hectares per year. According to GFMC daily updates, wildland fires continued to claim lives, destroy valuable private and public property and emit compounds that affect the composition and functioning of the atmosphere. Wildland fires and land-use fires consume an estimated

FIGURE 4
Area burned in five southern European countries, 1980–2003

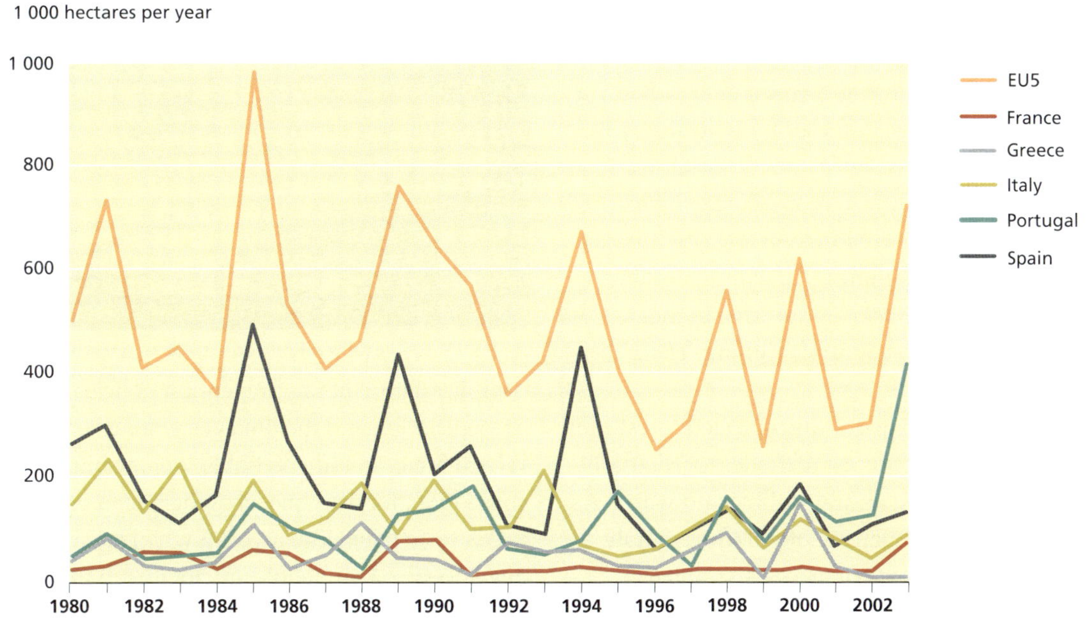

1 000 hectares per year

Source: UNECE/EC, 2004.
Note: Detailed statistics are only available for some regions, underscoring the need to improve coverage of satellite remote sensing systems.

average of more than 9 billion tonnes of vegetative biomass globally each year.

During 2002–2003, unprecedented high temperatures and drought in several regions broke records dating back 150 years. Extreme conditions resulted in severe fires in Australia (around Canberra), Canada (British Columbia), Italy, Portugal and the United States (California), causing the loss of more than 100 lives. Although fires in the United States in 2003 forced entire communities to be evacuated, the area burned (1.65 million hectares) was less than the average of the past eight years (2 million hectares) (NICC, 2003).

The number of fires and area burned fluctuate annually in the Mediterranean region. The total area burned in 2002–2003 did not exceed the area burned during extreme years of the 1980s and 1990s. In Portugal, fire area quadrupled compared with average years, and in France the area more than doubled the average (Figure 4). However, without the assistance of Italy and Spain, the figures would likely have been even higher. While additional funding and improved suppression technology will have an impact on the size of

a fire, more public education and awareness campaigns are needed to reduce the incidences.

In South Africa, large stores of industrial roundwood burned in 2003, while in the Russian Federation, 24 million hectares of coniferous forests and other lands were affected by wildfires in the same year, with devastating consequences for the ecology and national economy.

In tropical Asia and Latin America, land-use fires and associated smoke pollution continued to affect public health and safety, and the same problem appears to be surfacing in Central America. Through monitoring, GFMC has detected an increasing number of fire events in Central Africa, which indicates that in the equatorial forest region, fire is systematically being employed in land-use change, as in Asia and Latin America.

A number of wildfires throughout the world have led to secondary disasters of high humanitarian significance. Human casualties caused by mudslides occurring after fires or flash floods and public health affected by extreme wildland fire smoke pollution in many countries reveal that the consequences of

In 2003, 24 million hectares of coniferous forests burned in the Russian Federation

excessive burning and high-severity wildfires go beyond economic and biodiversity losses. Forest health is also affected by wildfires, which are often associated with insect infestation, for example of Siberian moth (*Dendrolimus superans sibiricus*) in Mongolia and the Russian Federation (Goldammer, 2004) and Southern pine beetle (*Dendroctonus frontalis*) in most Central American countries (Billings *et al.*, 2004).

The principal cause of uncontrolled forest fires in 2002–2003 was human activities, particularly from burning agricultural stubble and waste. As a case in point, 91 percent of wildfires in Italy originated from such practices. In Canberra, Australia, on the other hand, lightning caused the raging fires that destroyed 500 homes. Arson is on the rise, having been reported in Australia (Sydney), France, Mongolia, Portugal, the Russian Federation and the United States. Calls are therefore being made to tighten national fire legislation and to strengthen law enforcement.

Fire prevention

Fire prevention through sound management remains, by far, more cost-efficient than suppression during emergencies. However, lack of resources, negligence and policies focusing narrowly on conservation have left several areas without fire management strategies and increased their vulnerability. Prescribed burning and programmes to reduce fuel buildup are now priorities in Australia, Canada, the United States and elsewhere. In the United States, the main agencies involved in fire management (the Forest Service of the United States Department of Agriculture and the National Park Service and Bureau of Land Management of the Department of the Interior and others) conducted prescribed burning on more than 1 million hectares for fuel load reduction and other objectives (biodiversity conservation) in 2003.

The use of fire in agricultural practices in many developing countries illustrates the complex nature of fire prevention policies and fire legislation and the linkages among sectors. Where fire is an indispensable tool in shifting cultivation, for example, its widespread use needs to be taken into account when laws are developed so that people are not forced to break them to meet basic needs. Community forestry and similar programmes that engage residents in the quest for solutions have proved effective in both preventing and controlling wildland fires.

Fire suppression

While fire prevention may be the desirable approach, most countries pay dearly to maintain emergency response capacity to avoid severe social, economic and environmental losses. International cooperation, notably through bilateral agreements, is proving to be effective in combating fire and in facilitating emergency aid across borders. The introduction of an Incident Command System (ICS) – which provides a common language for international fire fighting teams in order to avoid misunderstandings in terminology – made it possible for fire fighters from Australia, Canada, Mexico, New Zealand and the United States to work together during 2002–2003 in Australia and the United States. This standardized system increases the safety of crews on the ground and in the air and decreases the risk of lives being lost.

International cooperation in wildland fire management

The global fire community met in 2003 at an International Wildland Fire Summit in Sydney,

Australia, to propose and agree on pragmatic and sustainable solutions to protect human health and avoid the consequences of wildland fires. More than 80 participants from 34 countries and 10 international organizations reached agreement on:

- principles to adapt international wildland fire management projects and exchanges to local ecological and social conditions;
- a template for international agreements that agencies can use to cooperate or to arrange mutual aid with one or more countries;
- the establishment of an Incident Command System as an international communication standard for wildland incident management;
- a strategy for enhancing future international cooperation in wildland fire management;
- a request for the United Nations (UN) to assist in implementing proposed strategic goals.

The need for countries to enter into collaborative agreements to assist with fire emergencies is recognized and well articulated. Indeed, commitment to move in this direction is illustrated by the results of the fire summit, the UN-led Inter-Agency Task Force for Disaster Reduction, the Global Fire Partnership that IUCN, The Nature Conservancy and WWF launched in 2003 and the establishment of 12 Regional Wildland Fire Networks within the Global Wildland Fire Network. The agreement that GFMC, the International Strategy for Disaster Reduction, FAO and the Global Observation of Forest and Land Cover Dynamics reached in May 2004 on a framework for developing an international wildland fire accord is further proof of fruitful collaboration.

Although the responsibility for suppressing fires resides with countries and national fire authorities, the key to dealing more effectively with emergencies lies in putting agreements in place between and among countries. To enhance this type of collaboration, FAO and partners are working with countries to develop bilateral or multilateral instruments.

A Workshop on Multilateral Assistance against Forest Fires in the Mediterranean Basin was held in Zaragoza, Spain, from 10 to 11 June 2003 under the auspices of the African Forestry and Wildlife Commission/European Forestry Commission/Near East Forestry Commission (AFWC/EFC/NEFC) Committee on Mediterranean Forestry Questions *Silva Mediterranea*. Participants studied procedures to coordinate mutual agreements and examined common legal and logistical tools to facilitate the sharing of resources among countries to combat forest fires within the Mediterranean Basin, when needed. This workshop was a preliminary activity to prepare a future Mediterranean conference on multilateral assistance against forest fires.

In April 2004, fire brigades of several European Union (EU) countries (including France, Germany, Italy, Slovenia and Spain) jointly conducted a large fire suppression exercise in southern France involving aerial means and ground crews. In the same year, consultations on cooperation in wildland fire management were held for the Balkan countries, the eastern Mediterranean, the Near East and Central Asia; the Baltic countries; Central America and the Caribbean; Northeast Asia; South America; the Southern African Development Community (SADC) and sub-Saharan Africa; and the Western Hemisphere. ◆

REFERENCES

Ahern, F., Goldammer, J.G. & Justice, C., eds. 2001. *Global and regional vegetation fire monitoring from space: planning a coordinated international effort.* The Hague, SPB Academic Publishing bv.

Ananthaswamy, A. 2004. Massive growth of ecotourism worries biologists. *New Scientist,* 4 March 2004 (available at www.newscientist.com/news/news.jsp?id=ns99994733).

Barrow, E., Timmer, D., White, S. & Maginnis, S. 2002. *Forest landscape restoration: building assets for people and nature – experience from East Africa.* Cambridge, UK, World Conservation Union.

Billings, R.F., Clarke, S.R., Espino Mendoza, V., Cordón Cabrera, P., Melendez Figueroa, B., Ramón Campos, J. & Baeza, G. 2004. Bark beetle outbreaks and fire: a devastating combination for Central America's pine forests. *Unasylva,* 217: 15–21

Institutional issues

Internal and external factors such as public pressure and economic realities are continuing to influence change in the forest sector and to shape the way forestry is defined and practised. Policies in other natural resource sectors are having a direct impact on sustainable forest management, increasing the urgency to improve synergies and strengthen partnerships. The recent expansion of the EU will also bring about new challenges and opportunities, also influencing markets for forest products. This chapter notes the latest trends in privatization; shows how modern reforms, including new technologies, are affecting the ways in which forests are managed; identifies forces driving forestry in countries with economies in transition; updates progress in forest law compliance; and outlines some of the challenges that developed countries must face in measuring and reporting their use of forests and wood products to meet commitments under UNFCCC and the Kyoto Protocol.

TRENDS IN PRIVATIZATION IN THE FOREST SECTOR

Governments have often used privatization measures to improve economic performance, especially since the end of the 1970s. Between 1985 and 1999, more than 8 000 transactions of this nature were completed around the world, for a total value exceeding US$1.1 trillion (in constant 1985 US dollars) (Brune, 2004). From the sale of state-owned enterprises alone, the Organisation for Economic Co-operation and

FIGURE 5
Amounts raised from privatization in OECD countries, 1990–2001

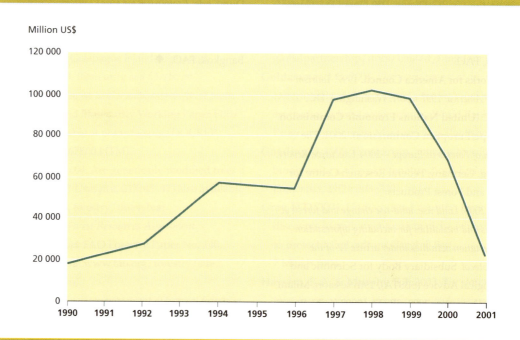

Source: OECD, 2002.

Development (OECD) countries received about US$693 billion between 1990 and 2001 (Figure 5).

Forests, however, were not among the first assets to be privatized, partly because of the sensitivities surrounding sovereignty, a growing recognition of their importance in protecting the environment and in providing services to society, and perceived high risks or low returns. Rather, initial efforts to privatize focused more on goods and services that brought a better return on investment, showed clear market opportunity and were less prone to civil society opposition. As it stands, privatization in the forest sector usually entails the transfer of property rights through the sale of natural forests or planted forests and through the devolution of forested land. Governments also involve the private sector through lease or concession contracts and outsourcing for services.

In the 1970s and 1980s, only a limited number of countries privatized forests. Chile moved in this direction, laying the foundation for a rapidly growing plantation industry; the Forestry Commission in the United Kingdom sold a small portion of its forest area; and China began to transfer rights associated with use and management in many parts of the country. In the 1990s, water, land and forests were more frequent targets for privatization, since few options were available in many countries. In 1999, privatization of primary industries such as petroleum, mining, agriculture and forestry overtook privatization of infrastructures.

Planted forests
Since 1974, the use of government incentives in Chile has resulted in the expansion of private planted forests to more than 2 million hectares. In New Zealand, privatization started in the late 1980s with the sale of 550 000 ha of state-owned forests, sawmills, nurseries and other assets. The sell-off of long-term cutting and management rights to domestic and foreign investors followed in the early 1990s. By 2000, 94 percent of planted forests in New Zealand were privately owned, but not the land (Ministry of Agriculture and Forestry, New Zealand, 2002). Similarly, South Africa privatized an estimated

90 000 ha of planted forests between 2000 and 2002, and the process is continuing (H. Koetze, personal communication, 2004). Several other countries in Africa are also taking measures to privatize planted forest, for example Ethiopia, Ghana, Kenya, Malawi, Mozambique, Nigeria, Uganda, the United Republic of Tanzania, Zambia and Zimbabwe.

Protected forest areas
Increasingly, private entities and NGOs are purchasing forest areas and acquiring land through concession contracts for protection and conservation purposes. For example, 32 percent of the national park area in Lithuania and 50 percent of protected forest areas in the Czech Republic are privately owned (Indufor and EFI, 2003). In Chile, private-sector interest in managing forests for conservation is growing as well. Governments in several countries, including Canada and the United States, are also discussing the possibility of outsourcing the management of protected areas.

Natural forests and woodlots
The privatization of natural forests through the transfer of land or forest ownership is less marked than that of planted forests, except in Central and Eastern Europe, where forest land is being returned to former owners. Trends vary across regions, depending on the economic model and the social and environmental conditions.

More common forms of private-sector involvement in the management of natural forests are concessions or leases, volume permits or standing timber sales, outsourcing and community-based approaches. According to a conservative estimate, the proportion of forests owned or administered by communities has doubled in the past 15 years, to about 350 million hectares (Scherr, White and Kaimowitz, 2003).

Regional trends
Africa. In most African countries, the state owns forest resources and allocates use rights through administrative or competitive mechanisms. In Gabon, 221 forest concessions are managing

11.9 million hectares, or 56 percent, of the forest area (Global Forest Watch, 2000). Cameroon has allocated 81 percent of its forests to concessions, of which 37 percent have been granted (White and Martin, 2002). Because of the importance of market forces and privatization in forestry, governments are reforming policies to be in a better position to move towards sustainable development. In some areas, however, armed conflicts prevent or slow down private sector involvement in the sector (see page 116).

In South Africa, the private sector owns and manages 70 percent of plantations (GCIS, 2004) – a trend that generally characterizes the current and possible future role of private companies in industrial plantations and outgrower schemes in southern Africa.

Impacts of European Union expansion on markets for forest products

Ten new members joining the EU means an increase in the trading bloc's population by 20 percent to 454 million. The establishment of a larger internal market should encourage trade and thus help to improve economies and raise standards of living. Larger EU membership could benefit the forest sector through:

- savings in transport time because of open borders;
- freer movement of labour;
- consistent quality control and trade regulations;
- better market information;
- almost 25 percent more forests available for wood supply.

Source: UNECE/FAO, 2004.

Asia. Privatization of the forest sector in Asia involves both entrepreneurs and communities. Participation of the latter is increasing as they are given the right to manage forests close to their villages through project-based activities and joint schemes.

After 1997, Malaysia guaranteed secure tenure for 100 years to private companies in Sabah through agreements covering more than 2.5 million hectares. In 2000, some 650 concessions of 69 million hectares were reportedly granted in Indonesia, although fewer than half were operational by the end of the year, covering almost 34 million hectares (Matthews, 2002).

Since the early 1980s, China has encouraged private investment by devolving use and management rights to households while retaining ownership of forest land. Both Chinese and foreign-owned companies are now making deals with communities and with households, to mutual benefit.

In India, 63 600 communities participate in joint forest management and are protecting and regenerating almost 14 million hectares or more than 19 percent of the forest land (Press Information Bureau, Government of India, 2003).

Central and Eastern Europe. In most Central and Eastern European countries, property expropriated by former regimes is being returned to owners, including forest land. Institutional restructuring and the rapid development of forest-based industries are also taking place. Many of the more than 4 million new forest owners, holding an average of about 2 ha, are inexperienced in forest management, business or market economies.

In countries that have joined the EU (the Czech Republic, Estonia, Hungary, Latvia, Lithuania, Poland, Slovakia and Slovenia) or are about to join (Bulgaria and Romania), restitution efforts cover 2.8 million hectares of forests, with the state still owning 63 percent of total forest area.

By June 2003, more than 1.4 million hectares, or 29 percent, of forests in Romania had been returned to former owners, mostly to municipalities and communities. Individuals received slightly more than 224 000 ha. National

In Slovakia, as in most Central and Eastern European countries, forest land expropriated by former regimes is being returned to owners

forest administrations at the central and branch levels targeted the end of 2004 for completing the process (Indufor and EFI, 2003).

Latin America. In Latin America, the state owns most natural forests and, in many countries, the transfer of land titles is limited to poor farmers. Some forest areas have been allocated to private land use as a result of pressure from ranchers or private business. In the largest tracts of natural forests, land-use change is intimately linked to agricultural demands.

In Peru, following the enactment of legislation pertaining to forests and wildlife in 2002, the government assigned 21 million of 67.5 million hectares of forests for timber production through concessions covering between 5 000 and 40 000 ha for up to 40 years (*El Peruano* newspaper, 2002).

In Bolivia in 2003, 5.4 million hectares, or 10.2 percent, of the forest area was managed as regular concessions. The government awards different types of land leases for long-term contracts (400 000 ha) and scientific research (200 000 ha) (Scherr, White and Kaimowitz, 2003).

Ecuador outsources forest administration, while in the Dominican Republic, independent foresters monitor the implementation of government-approved forest management plans on private land and report findings to authorities.

Commonwealth of Independent States. While these countries have not yet transferred ownership of forest resources, involvement of the private sector is increasing, mainly through the transfer of long-term use rights in the form of forest concessions.

In the Russian Federation, forest resources are likely to remain under state ownership, but the private sector is becoming involved through concessions or other contractual arrangements supervised by the forest administration. It is expected that most production forests will be managed this way, with the state retaining authority over conservation.

TRENDS IN FORESTRY ADMINISTRATION

Modern reforms are opening opportunities and creating challenges for forestry administrations around the world. Drivers for change include the transition from command and control to market economies; sustainable development; globalization; political, economic and social-equity dimensions of governance; and new technologies, including information technology.

Functions and methods of operation

In response to public demand for greater accountability, increased participation in planning and decision-making and better delivery of goods and services, central forestry

The most common methods to privatize the forest sector in the past three decades have varied according to the economic model, type of forest resources and desired outcome.

- Transfer of property rights is completed either through the sale of forest resources to the highest bidder or to a preferred beneficiary with or without a financial transaction. This method was widely used to privatize planted forests in some countries, for example New Zealand and South Africa.
- Restitution involves governments' handing back productive assets to former owners through transfer of resource tenure, revenue ownership and management rights to individuals or corporate bodies. This method is used in Central and Eastern Europe and to a limited extent in South Africa.
- Transfer of use rights to private companies, communities or households maintains government ownership of the forest resource. It involves either an administrative allocation of resources or a lease through competitive tenders. Governments may retain rights to decide the flow of goods and services.
- Procurement or outsourcing of private-sector services maintains government ownership and responsibility for deciding the flow and distribution of goods and services. Governments may outsource forest management and operational activities such as inventory, harvesting, silviculture and forest protection.

administrations are increasingly transferring resources and responsibilities to subnational governments and are delegating more functions down the chain of command. In general, policy and regulatory functions remain with central governments, while the private sector and civil society are taking charge of operations.

Decentralization has also resulted in a larger role for municipalities.

In Africa, Asia and Latin America, more than 30 countries report some degree of decentralization in the sector. Faced with limited and sometimes dwindling resources, forestry administrations are also reducing personnel and streamlining operations. For example, Argentina, Costa Rica, New Zealand and South Africa have introduced reforms in an effort to cut costs and increase efficiency.

Organizational structures

Many forestry administrations that have national responsibility for commercial production, conservation and extension are moving towards a three-tier system: a national component with reduced personnel to direct and implement change; regional units to coordinate activities and give technical guidance; and municipal and local units to manage the resource. Chile, Costa Rica and the Sudan, among others, are structured along these lines.

National component. At the national level, a streamlined central unit fulfils state responsibilities for the stewardship of natural resources, for strategic planning and coordination and for the provision of public goods. This unit also develops and analyses policies, establishes national goals and directs the collection and management of information for decision-making. The establishment of self-financing units is one of the new approaches being used to carry out such functions and to overcome salary limitations that often prevent public services from hiring the most qualified professionals. Examples of smaller, less costly organizations that are financed with revenues from forest fees and fines are found in the Sudan and Suriname.

Regional units. Structured along the lines of central administrations, regional units provide technical guidance at the local level when capacity is insufficient, for example, to develop forest management systems, prevent and control forest fires and address issues related to forest health. In production forests, these units can also develop guidelines for silviculture and for biodiversity conservation. Officials work with other regions and municipalities to coordinate initiatives, collect information and monitor activities.

Municipal and local levels. Subnational governments and stakeholders close to the forest who possess knowledge of the resource and of local customs, demands and values are becoming key players in forestry administrations and in forest management, especially in sub-Saharan Africa and on community-owned lands in Asia. In countries where land-use rights are more formally defined, farmer associations and small entrepreneurs mostly assume this responsibility, subject to municipal regulations. In general, responsibilities for sustainable forest management on the ground rest at the municipal level. In addition, administrators at this level resolve local conflicts related to forests and promote public participation in planning and decision-making.

Advances in technology

Technological changes are opening important opportunities to improve ways in which the sector is governed and how administrations operate. Advances in information and communication technologies, including satellite imagery and detection as well as spatial information and decision-support systems, offer the greatest potential to achieve gains.

Given that many forestry administrations have invested in information technology such as digital cartography, planning and policy analysis are expected to improve significantly as a result of better data and more extensive databases. Such developments should enhance the effectiveness of national forest programmes and foster greater participation and transparency in forestry administrations.

Spatial information technologies and related satellite applications have opened the way for multipurpose information systems and enhanced the capacity of forestry administrations. Many countries, for example, are using satellite technology to detect

---

National forest programmes

A national forest programme is both a dynamic process responsive to change and a framework for planning and action. It provides strategic orientation to the forest sector and facilitates coordinated implementation of sustainable forest management. Basic characteristics include:

- national sovereignty and country leadership;
- consistency with national constitutional and legal frameworks;
- linkages to national sustainable development strategies;
- complementarity to international agreements relevant to the forest sector;
- approaches that integrate the range of values and functions of forests and trees;
- cooperation and collaboration across sectors;
- partnerships;
- participatory policy development, planning, implementation and monitoring.

Since FAO established an online information platform on national forest programmes in 2003, more than 90 countries have prepared profiles that are being made available online. In addition, the Organization is supporting 22 member countries with implementation efforts.

To assist developing countries with national forest programmes, a number of international organizations and donors, including FAO and the National Forest Programme Facility, are helping to link the programmes with broader agendas, address governance issues, develop national capacity and make knowledge available to those involved in the process.

Changes in forest management in transition economies

In February 2003, a workshop held by the Ministry of Natural Resources of the Russian Federation and the World Bank, with support from the Program on Forests (PROFOR), identified the following factors driving reforms in countries with transition economies:

- dramatic changes in the business environment in the past decade;
- continuing unfavourable investment climate for developing forest industries;
- flexible tenure systems that are site and situation specific;
- appropriate rent capture/taxation and sound financing of forest management;
- forest certification to secure environmentally and socially sensitive markets;
- institutional changes to respond to the needs of a market economy and competition.

The workshop, which was held in Moscow, the Russian Federation, and attended by almost 100 experts, pointed out that large-scale projects need to be flexible enough to respond to sometimes rapidly changing policy and legal environments; and that in larger countries, diverse geographic and socio-economic conditions need to be taken into account before a particular approach to institutional change is chosen (PROFOR, 2003).

forest fires and to help assess the extent of deforestation and forest degradation. Others are using satellite technology for monitoring and planning purposes.

Government organizations in other sectors are applying these technologies as well, especially with regard to land use. The development and modernization of land administration systems will have significant bearing on information regarding land rights, responsibilities and restrictions, and will facilitate land transactions around the world. The ease with which land can be bought and sold as a result of reliable information on ownership means that the forest sector could become more open to free trade and globalization. In this new era, forestry administrations will be expected to focus on policy development and move away from traditional operations.

Constraints and opportunities

Streamlined forestry administrations based on a central system, regional networks and local participation are changing the ways in which forests are managed. Information technology increases capacity for planning, monitoring and assessment, and facilitates wider involvement of parties. However, the chronic lack of resources and low public investment in forestry remain problematic.

Reforms are affecting administrations in other natural resource sectors as well, increasing the urgency to establish synergies and partnerships. Although changes in forestry administrations are taking place, they do not appear as extensive as in other areas. For example, governments are creating new executive agencies to deal with territorial administration and decentralization, local government, capacity building and gender development. These new entities perform functions that are closely related to forestry and sometimes assume responsibilities that once belonged to others, making it all the more important to share information, coordinate activities and invest in managing change.

Forestry administrations must prepare staff to deal with new realities, apply and master emerging technologies and take steps to ensure that all levels of authority have access to the knowledge and skills they require to perform their tasks.

EFFORTS TO IMPROVE FOREST LAW COMPLIANCE

Governments, with the help of international organizations, NGOs and the private sector, are continuing their efforts to improve forest law compliance. Most initiatives are built on the premise that, although important, compliance

Decentralization and national forest programmes

In April 2004, 182 experts from countries and organizations around the world met in Interlaken, Switzerland, to share experiences on decentralizing forestry systems and to identify strategies that would allow national forest programmes to address issues related to the process. The Workshop on Decentralization, Federal Systems in Forestry and National Forest Programmes, held by the Governments of Indonesia and Switzerland in support of the United Nations Forum on Forests (UNFF), noted that decentralization is a means to alleviate poverty, achieve sustainable development and protect forest values. The experts recognized that the process is complex and dynamic and needs to take into account the specific conditions of each country. While progress is being made, participants also felt that true democratic decentralization has rarely been implemented to date as much of the decision-making, resources and benefits from forests remain with central authorities.

The workshop highlighted the need for the following actions, among others:
- developing a common understanding of concepts, terms and definitions related to decentralization in the forest sector;
- enhancing understanding of decentralization through information dissemination;
- formulating approaches to maintain protected areas while promoting the use of traditional knowledge and practices;
- developing principles to devise equitable representation and to devolve authority and resources for forest management to the lowest appropriate level;
- promoting the valuation of, and compensation for, the environmental services that forests provide;
- sharing information and establishing partnerships across sectors;
- integrating decentralization into national forest programmes at the national or subnational levels;
- strengthening the human and institutional capacity of stakeholders and promoting partnerships;
- involving NGOs and other major groups in the planning, implementation and monitoring of decentralization activities.

strategies can no longer rely on policing alone but must include efforts to streamline policy and legal frameworks; to provide incentives to comply with regulations; to improve employment conditions of enforcement officers; to conduct public education and awareness programmes; and to use national and international market restrictions to limit opportunities for trading illegally sourced wood. This section describes major undertakings to date.

Multilateral initiatives
As the need to improve forest law compliance has gained prominence in international discussions, the need to take concerted action to conserve and use species in a sustainable

manner has become increasingly apparent. The expanded Programme of Work on Forest Biological Diversity of CBD also includes actions to promote forest law compliance and address trade issues.

In 2001 and 2002, the United Nations Security Council investigated the role of illegal exploitation and trade of natural resources in fuelling the civil war in Liberia and, as a result, imposed an embargo on exports, transportation and imports of Liberian wood in 2003. In addition, G8 countries (Canada, France, Germany, Italy, Japan, the Russian Federation, the United Kingdom and the United States) strengthened their resolve to combat illegal activities in the forest sector, and in 2003 committed to supporting efforts in Africa.

UNFF is currently discussing issues related to illegal logging and trade as well, urging countries to improve law enforcement in the forest sector and to control illegal trade in forest products. It has also called on the international community to assist countries in building their capacity to improve forest law enforcement.

Following the Forest Law Enforcement and Governance (FLEG) East Asia Ministerial Conference in September 2001, a regional task force was established to identify ways to implement the declaration that was adopted during the meeting. As a result, Indonesia, for example, has entered into a partnership with the World Bank and WWF to develop a strategy that identifies action to be taken to enforce and prevent illegal acts in the sector.

The Asia Forest Partnership, launched at the World Summit on Sustainable Development (WSSD) in 2002, recognizes that many initiatives support sustainable forest management and the control of illegal forest activities in Asia, and aims to promote further cooperation in addressing urgent issues. Although the partnership does not focus exclusively on the control of illegal logging and forest law enforcement, both figure prominently among its objectives.

In May 2003, the European Commission unveiled an EU Action Plan for Forest Law Enforcement, Governance and Trade (FLEGT). Measures include support for improved governance in producer countries; partnerships with producer countries to ensure that only legally harvested timber enters the EU market; and international collaboration to combat trade in illegally harvested timber. Through the plan, the EU will help interested producer countries set up a voluntary scheme of licences to verify the legal origin of forest products before exporting them to its member countries. The EU also supports activities to restrict investments that may induce illegal transactions and is addressing the use of illegally sourced forest funds to finance armed conflicts. Along with FLEG, the plan is one of the most comprehensive to fight illegal logging and associated trade.

In October 2003, under the auspices of the New Partnership for Africa's Development (NEPAD),

African ministers pledged to fight violations to forest law by strengthening national initiatives and collaborating on a bilateral, regional and multilateral basis. Their declaration outlines 38 actions that countries should take to improve law enforcement in the region. NEPAD's efforts complement other initiatives to bring about change in natural resource management such as the Congo Basin Forest Partnership.

In 2003, the Ministerial Conference on the Protection of Forests in Europe (MCPFE) signed the Vienna Living Forest Summit Declaration, in which parties commit, among other actions, to improving governance in the forest sector, promoting the enforcement of forest laws, combating illegal harvesting of forest products and related trade and fostering sustainable forest management in Europe and elsewhere. A work programme is being developed to achieve these ends.

Agreements targeting illegal logging and illegal trade

As an example of exporter and importer countries working together to combat illegal logging and related trade, the Governments of Indonesia and the United Kingdom of Great Britain and Northern Ireland signed a Memorandum of Understanding in 2002. The countries made a commitment to develop systems of verification and compliance; increase the involvement of civil society; strengthen institutions, data collection and collaboration; and enlist the support of the private sector. Indonesia has also signed bilateral agreements with China, Japan, Malaysia and Norway to curb illegal logging and trade of Indonesian timber.

In July 2003, the United States launched the President's Initiative Against Illegal Logging, which focuses on three regions: the Amazon Basin and Central America; the Congo Basin; and South and Southeast Asia. The scheme supports activities related to good governance, community-based actions, technology transfer and optimum use of market forces. The United States has also launched an initiative in Liberia to stop illegal harvesting and restore deforested areas.

Work of international agencies and other organizations

FAO, in partnership with ITTO, has identified best practices and developed compliance guidelines to help decision-makers design and implement effective policies, legislation and institutional frameworks. FAO has also compiled a catalogue of national forest laws and conducted case studies to determine factors that facilitate or compel people to engage in illegal actions in the sector. The studies are providing insight into the causes of unlawful acts and potential remedial measures. FAO is also examining ways in which private forest corporations can adhere more fully to the law in the countries in which they operate by adopting codes of conduct.

ITTO is assessing the consistency of export and import data on tropical timber and related products and is continuing to assist member countries in designing frameworks for forest law enforcement. In partnership with WWF and others, it finalized a study on the potential role of phased approaches to timber certification, an important step in verifying the legality of wood being traded. In addition, ITTO, in cooperation with FAO, facilitated a meeting of major national and international forest certification schemes in June 2003 to increase mutual understanding of different approaches.

CIFOR examined the impact of law enforcement on rural livelihoods, analysing the situation in six countries in Africa, Asia, Latin America and North America. The initiative looked at ways to involve rural communities in reforms, increase awareness, identify gaps in knowledge and help develop strategies that address livelihood issues. CIFOR is also carrying out research on ways to use money-laundering legislation to curb illegal logging and signed a memorandum of understanding with the Government of Indonesia to develop measures to reduce money laundering linked to forestry crimes. In this regard, Indonesia became the first country to list forestry crimes as a predicate offence in its new money-laundering law.

The new forest strategy of the World Bank includes provisions to address corruption and illegal activities through better forest laws, regulations and enforcement. As part of its programme on governance in the sector, the World Bank supported the FLEG process and hosted a forum on forest investment with senior executives of forest companies, private- and public-sector financial institutions and leading conservation agencies from around the world. The forum concluded with a call to curb illegal logging and promote responsible investment. In addition, current and proposed policies of the World Bank, the African Development Bank, the Asian Development Bank and the Inter-American Development Bank all contain references to curbing illegal activities in the forest sector.

Various corporations are adopting codes of conduct, most of which include reference to illegal forest activities. The European Foundation for the Preservation of African Forest Resources, whose members include companies with concessions in Africa, made significant commitments to improve forest management in countries of the region. The Interafrican Forest Industries Association (IFIA) developed a code of conduct for members operating in the Congo Basin and humid West Africa. Other entities fighting illegal acts include the Japanese Federation of Wood Industry Associations, the only organization representing the country's wood industries; the International Council of Forest and Paper Associations, representing industries from 43 countries, 75 percent of the world's paper and more than 50 percent of the world's wood production; the Timber Trade Federation of timber importers from the United Kingdom; the International Technical Tropical Timber Association; the Confederation of European Paper Industries; and the American Forest and Paper Association. Individual corporations are also taking steps to avoid buying and selling illegally sourced timber.

In addition, NGOs such as Greenpeace International, the Environmental Investigation Agency, the World Rainforest Movement, Global Witness, Friends of the Earth International and Transparency International are working on

their own or with a number of governments to expose, monitor and help suppress illegal forest activities through education campaigns, studies and research.

CREATING NATIONAL FRAMEWORKS FOR FORESTS UNDER THE KYOTO PROTOCOL: CHALLENGES AHEAD

Almost three decades ago, Dyson (1977) proposed that harmful emissions of carbon dioxide (CO_2), the main cause of global warming, could be turned into new forests via the process of photosynthesis, thereby replacing some of the 16 million hectares of natural forests that the planet loses annually (FAO, 2001). Finally, 188 Parties to UNFCCC have elaborated rules and guidelines to put his idea into practice through the Kyoto Protocol.

Negotiating the extent to which industrial countries could use forests and wood products to meet their commitments to mitigate climate change proved time consuming and contentious. Rules are complicated, and measurement and reporting procedures are costly to the point that they may prevent some countries from using the full range of forestry activities eligible under the Kyoto Protocol. Now, countries face the formidable challenge of creating national frameworks to implement commitments in the context of their domestic forests, and little time remains until 2008, the start of the first commitment period. Three major tasks – acting on the general commitments, monitoring and reporting forest carbon stock changes and implementing the international climate change agreements – lie ahead:

- General commitments can be acted on with relative ease, for example, by including forests in national adaptation and mitigation programmes, raising awareness of the role of forests in climate change, promoting sustainable forest management and conserving and enhancing forest sinks.
- Monitoring and reporting forest carbon stock changes places demands on countries to develop methods for including carbon in forest inventories, in measurement protocols and in data management systems. In some

instances, meeting this requirement may require new laws and more reliable forest inventories.
- Implementing the international climate change agreements after ratification of the Kyoto Protocol will require new or revised legislation on forests and in other related areas at the national or subnational level, along with appropriate institutions to support implementation. Few countries have begun to tackle this aspect. Ownership of carbon in forests, trees and wood products is one of the key issues.

Who owns the carbon?
Carbon ownership comes with rewards but also with risks. In countries with ambitious afforestation and reforestation programmes, young, fast-growing forests can offset a substantial part of industrial CO_2 emissions to help fulfil reduction obligations (see Box on facing page). These new forests remove carbon from the atmosphere and reduce the need for a country to lower industrial emissions or purchase carbon credits to meet commitments. The question is whether private, community and subnational forest owners should undertake these activities without reward, particularly when fossil fuel emissions contain not only CO_2 but also sulphur, nitrogen and heavy metals, which, as components of acid rain, harm their forests.

The risky side of owning carbon rights is linked to the obligation for countries to account for carbon released during the commitment period as a result of all deforestation since 1990. Should a private forest owner, after converting a forest to pasture, be liable for the carbon released from trees, soils and litter during the first and possibly subsequent commitment periods? Or should the government, ultimately responsible under the Kyoto Protocol, assume ownership of, and liability for, all gains and losses from afforestation, reforestation and deforestation?

With regard to afforestation, reforestation and deforestation since 1990, industrialized countries are obliged to account for the net carbon stock changes that result from these actions. For forests established before 1990, they may opt for forest management, as defined under the Kyoto Protocol, as one of several eligible activities. If carbon stocks in these older forests increase, a country may gain credits up to a specified upper limit. On the other hand, a country also risks incurring debits if domestic growing stocks decrease as a result of accelerated harvesting, for example.

Again, the ownership question arises. Should the government avail itself of sequestration in older domestic forests without compensating owners? Should owners receive payment in proportion to the growing stock increment in their forests? In turn, are owners prepared to risk losses or to pay back carbon revenue after harvest? Should owners be eligible to sell carbon fixed by their forests in domestic or even in regional or international markets?

For most industrial countries, credit allowances for forest management amount to only 15 percent of the total carbon increment of domestic forests. Governments will need to decide whether they will draw exclusively on state-owned forests to fill the national quota, thereby possibly creating a disadvantage for private forest owners and a distortion in the timber market; whether they will award credits only to those who take deliberate action to enhance carbon sequestration in their forests; and which forest management practices should be recognized to achieve such results.

Developing countries do not have quantitative greenhouse gas reduction commitments. In the context of CDM, the host country must nevertheless recognize that foreign investors in afforestation and reforestation projects have rights to all or part of the carbon sequestered by CDM projects or that ownership of sequestered carbon may be transferred abroad, independent of ownership of the timber.

Giving forest owners the rights to sequestered carbon involves additional issues (FAO, 2004), such as how to:

- assess, verify and record sequestered carbon;
- promote orderly sales or other transfer of ownership;

- allocate the risk of failure of carbon sequestration;
- assess liability for damages to, or elimination of, the potential of a forest to sequester carbon.

National legal and policy frameworks

Beyond clarifying ownership rights, countries may enhance net carbon sequestration in forests by other means (see Box on page 54). Approaches could encompass laws that restrict harvesting, harvest methods and ages, silvicultural systems, treatment of logging slash, regeneration lag, minimal stocking, fire

Valuing carbon sequestration in Irish forests

Ireland's industrial emissions will probably exceed Kyoto commitments, which entail annual emission reductions of approximately 15.4 million tonnes of CO_2 or 4.2 million tonnes of carbon (Bacon, 2003). Forests established since 1990 will fix 0.3 million tonnes of carbon per annum, offsetting about 6.5 percent of Ireland's projected excess emissions and reducing carbon credits to be acquired in international markets by this amount. At a market value of €30 per tonne of carbon in international emission trading, these young Irish forests alone would save the country an expense of about €9 million annually or €45 million over the commitment period 2008–2012.

The average rate of carbon gain in these young forests is estimated at 3.4 tonnes of carbon per hectare per year. They would thus accumulate a carbon value of approximately €100 per hectare annually.

Credits for forest management in Ireland are capped at 50 000 tonnes of carbon per year. If the country chooses forest management as an eligible activity under the Kyoto Protocol, an additional value of €1.5 million could accrue annually in the form of carbon revenue.

protection and controlled burning. Where forest management agreements or concessions regulate forestry operations, applicable laws and contracts may need to be revised.

In some instances, laws might need to be streamlined to facilitate climate change mitigation projects. Carbon sequestration projects in California, for example, were subject to at least 16 federal and state regulations (Vine, 2004). The Kyoto Protocol and many countries require environmental and social-impact assessments for afforestation and reforestation (Bekhechi and Mercier, 2002). The carbon sequestration services that forests provide should probably be given a weight in these assessments as well as in laws on land-use planning or zoning (Kennett, 2002). In some countries, laws on landscape conservation require material offsets for human interventions. In Germany, for example, the administration responsible for constructing a new highway through forest lands must compensate for lost forest services by establishing new forests or by enhancing biological diversity or other services in adjacent forests.

Countries might also enhance carbon fixation by forests through subsidies, taxes, risk reduction, research, extension services and public awareness initiatives. Moreover, national forest programmes appear to be an effective means to integrate the opportunities, rules and modalities of the Kyoto Protocol into national forest policy and planning.

Future challenges

National frameworks under the Kyoto Protocol hinge on institutional capacities and on countries identifying a Designated National Authority if they are contemplating using CDM. At last count, only the European Community, eight industrial countries, 39 developing countries and six countries with economies in transition had done so. Since 2002, FAO has helped build capacity for CDM in Central America and, with IUCN and the United Nations Environment Programme (UNEP), in Africa, Asia and Latin America.

Parties to UNFCCC have now established most modalities, rules and guidelines at the international level. As the first commitment period approaches, only a few countries have decided if and how to use their forests for climate change mitigation and adaptation. Consequently, little effort has gone into developing national legal and institutional frameworks for implementing the protocol in the forest sector. Many challenges lie ahead in this regard. ◆

Creating a domestic framework for forests and climate change in Spain

Spain adopted a new forest plan in 2002 and a national forest law in 2003 that define domestic forest policy regarding climate change. The plan considers policy support crucial to climate change mitigation. It establishes the potential for mitigation based on available area, evaluates technical capacity for sequestration and assesses possibilities for enhancement. The law recognizes global climate change mitigation and wood energy as valuable functions of forests that should be enhanced. Public administrations may grant subsidies, conclude contracts with owners or invest directly in public lands to achieve goals. Research on energy use from logging residues and adaptation of forests to climate change has also been initiated.

REFERENCES

Bacon, P. 2003. *Forestry: a growth industry in Ireland* (available at www.coford.ie/activities/BaconReport.pdf).

Bekhechi, M.A. & Mercier J.-R. 2002. *The legal and regulatory framework for environmental impact assessments*. Washington, DC, World Bank.

Brune, N. 2004. *Privatization around the world*. New Haven, USA, Yale University. (PhD thesis)

Dyson, F.J. 1977. Can we control carbon dioxide in the atmosphere? *Energy*, 2: 287–291.

El Peruano **newspaper.** 2002. Lima, 22 March 2002.

FAO. 2001. *Global Forest Resources Assessment 2000.* FAO Forestry Paper 140. Rome.

FAO. 2004. *Climate change and the forestry sector. Possible legislative responses for national and subnational governments.* (In press)

GCIS. 2004. *South Africa Yearbook 2002/03.* Pretoria, Government Communication and Information System (available at www.gcis.gov.za/docs/publications/yearbook.htm).

Global Forest Watch. 2000. *A first look at logging in Gabon.* Washington, DC, World Resources Institute (available at www.globalforestwatch.org/common/gabon/english/report.pdf).

Indufor Oy & European Forestry Institute (EFI). 2003. *Forestry in accession countries.* Final report prepared for the European Commission DG Environment. Helsinki.

Kennett, S.A. 2002. National policies for biosphere greenhouse gas management: issues and opportunities. *Environmental Management,* 30: 595–608.

Matthews, E., ed. 2002. *The state of the forest: Indonesia.* Bogor, Indonesia, Forest Watch Indonesia, and Washington, DC, Global Forest Watch (available at www.globalforestwatch.org/common/indonesia/sof.indonesia.english.low.pdf).

Ministry of Agriculture and Forestry, New Zealand. 2002. *A national exotic forest description as at 1 April 2001.* Wellington.

OECD. 2002. *Recent privatization trends in OECD countries.* Paris, Organisation for Economic Co-operation and Development (available at www.oecd.org/dataoecd/29/11/1939087.pdf).

Press Information Bureau, Government of India. 2003. "Forest area under peoples' management has doubled – joint forest management committees to be further strengthened". Press release, 10 February.

PROFOR. 2003. *Institutional changes in forest management – experiences of countries with transition economies: problems and solutions. Workshop Proceedings.* Washington, DC, Program on Forests (available at www.profor.info/pubs/governance.htm).

Scherr, S., White, A. & Kaimowitz, D. 2003. *A new agenda for forest conservation and poverty reduction. Making markets work for low-income producers.* Washington, DC, Forest Trends, and Bogor,

Indonesia, Center for International Forestry Research (CIFOR).

UNECE/FAO. 2004. *Forest Products Annual Market Review, 2003–2004.* Timber Bulletin LVII(3). Geneva, Switzerland, United Nations Economic Commission for Europe (available at www.unece.org/trade/timber/docs/fpama/2004/2004-fpamr.pdf).

Vine, E. 2004. Regulatory constraints to carbon sequestration in terrestrial ecosystems and geological formations. *Mitigation and Adaptation Strategies for Global Change,* 9: 77–95.

White, A. & Martin, A. 2002. *Who owns the world's forests? Forest tenure and public forests in transition.* Washington, DC, Forest Trends (available at www.forest-trends.org/resources/pdf/tenurereport_whoowns.pdf). ◆

International forest policy dialogue

Countries have been engaging in international forest policy dialogue in a number of fora since before UNCED. In some instances, forest issues are not arising from the discussions as a formal part of the agenda, yet decisions directly affect forest management and practices around the world. The Millennium Development Goals and the Plan of Implementation of the World Summit on Sustainable Development (WSSD), for example, recognize that forests are critical to achieving overall sustainable development, reducing poverty, improving the environment, halting the loss of biodiversity and reversing land and resource degradation. These significant contributions argue strongly for all sectors to adopt more comprehensive approaches in their search for lasting solutions. Any future international arrangements on forests, therefore, may need to reach out to experts in other fields to help shed light on issues of global concern that have been with the sector for some time. This chapter presents a historical perspective and the current status of the international forest policy dialogue; summarizes the results of the fourth session of UNFF; describes progress in the work of CPF; and gives an update on international conventions and agreements related to forests.

A HISTORICAL PERSPECTIVE

Countries have been discussing international forest policy issues within the United Nations system since the end of the Second World War. Since its establishment in 1945, FAO has incorporated forestry as one of its major programmes to address concerns that have gradually shifted from timber supply in post-War Europe to broader development and conservation issues. Since the late 1940s, six

Regional Forestry Commissions have been bringing the heads of national forest agencies together on a regular basis to discuss policy and technical matters. The Committee on Forestry (COFO) – the most important of the FAO statutory bodies in forestry – first met in 1972. Forestry is also discussed every two years at the Conference of FAO.

Concerned with the increasing rate of deforestation in tropical regions, the FAO Committee on Forest Development in the Tropics (1967–1991) brought worldwide attention to the problem. In 1983, this intergovernmental forum called for the development of a plan to reverse trends. In response, FAO, the World Bank, the United Nations Development Programme (UNDP) and the World Resources Institute (WRI) launched the Tropical Forestry Action Plan (TFAP) (later Tropical Forestry Action Programme) in 1985. However, TFAP became mired in controversy and was replaced with the National Forestry Action Plan (NFAP) (later National Forestry Action Programme), which then evolved into national forest programmes in the 1990s.

Other milestones in the global forest policy dialogue were the adoption of the International Tropical Timber Agreement (ITTA) in 1983 and the establishment of the International Tropical Timber Organization (ITTO) in 1986 to promote international trade in tropical timber, the sustainable management of tropical forests and the development of forest industries. Many regional initiatives to protect forests and promote their sustainable management have emerged as well: the Amazon Cooperation Treaty, established in 1978, and the Ministerial Conference on the Protection of Forests in Europe (MCPFE), established in 1990, for example.

From UNCED to IPF/IFF/UNFF

The United Nations Conference on Environment and Development (UNCED), held in Rio de Janeiro, Brazil, in 1992, marked a turning point in the international forest policy dialogue when countries affirmed their commitment to sustainable forest management by adopting the Non-Legally Binding Authoritative Statement of Principles for a Global Consensus on the Management, Conservation and Sustainable Development of All Types of Forests (the "Forest Principles"). This agreement provided, for the first time, a common basis for action at the national, regional and international levels. It was also significant for the compromise it represented at a time when countries could not reach consensus on whether to launch negotiations for a global forest convention.

To follow up the UNCED forest-related outcomes, the Economic and Social Council of the United Nations (ECOSOC) established the Ad Hoc Intergovernmental Panel on Forests (IPF) (1995–1997) and subsequently the Intergovernmental Forum on Forests (IFF) (1997–2000). Their mandates were to promote and facilitate implementation of the Forest Principles; to review progress towards sustainable forest management; and to seek consensus on future international arrangements. The dialogue resulted in nearly 300 Proposals for Action. However, in addition to the question of a legally binding instrument, thorny issues related to finance, transfer of technology and trade remained unresolved. The importance of these matters and the desire to keep forests on the global policy agenda led to the establishment of the United Nations Forum on Forests (UNFF) in 2000 for an initial five-year period.

UNFF not only provides a forum for sharing experiences and lessons on implementing sustainable forest management, including IPF/IFF Proposals for Action, but has also hosted a ministerial segment and organized multistakeholder dialogues. Several country-led initiatives have facilitated more in-depth deliberation on key issues, resulting in more informed discussions during formal sessions. Indications to date are that the dialogue seems

International Year of Freshwater

The United Nations General Assembly proclaimed 2003 the International Year of Freshwater to raise awareness of the need to use and manage this resource in a sustainable manner. It called upon governments, national and international organizations, NGOs and the private sector to make voluntary contributions and to lend other forms of support to the initiative and its messages. The designation provided an opportunity to accelerate implementation of the principles of integrated water resources management and to spearhead new initiatives at the international, regional and national levels.

Throughout the year, FAO highlighted the critical links between mountains, forests and freshwater. The Organization remains actively engaged in assisting countries around the world to address related issues, for example, by developing guidelines and approaches. As the agency mandated to lead observation of the first International Mountain Day, celebrated on 11 December 2003, FAO chose "Mountains – source of freshwater" as its theme.

to be effective and meaningful at the regional level and has served to strengthen political commitment among countries sharing similar conditions, such as those that are members of criteria and indicators processes and Regional Forestry Commissions.

Policy deliberations relevant to forests are also under way in a myriad of other international fora, mainly the Convention on Biological Diversity (CBD), the United Nations Framework Convention on Climate Change (UNFCCC) and the United Nations Convention to Combat Desertification (UNCCD) – all of which arose from UNCED. Efforts are being made to enhance cooperation on forest issues among these entities as well as other processes and organizations, particularly through the Collaborative Partnership on Forests (CPF).

UNITED NATIONS FORUM ON FORESTS

UNFF held its fourth session in May 2004 in Geneva, Switzerland. Governments discussed social and cultural aspects of forests; traditional and scientific forest-related knowledge; and progress in achieving sustainable forest management, through *inter alia* implementing the IPF/IFF Proposals for Action and criteria and indicators. Delegates also addressed the need to enhance cooperation among international organizations and, in this regard, expressed appreciation to CPF for its work. In addition, governments established procedures for reviewing the international arrangement on forests, including UNFF and CPF, in 2005. Panel discussions also took place on the role of forests in rural development and poverty eradication in Africa and SIDS, and in achieving the United Nations Millennium Development Goals.

Among the outcomes of UNFF-4 were the acknowledgement of seven thematic elements of sustainable forest management (see Box on page 3) and the need to strengthen linkages between forests and internationally agreed development goals. Member countries,

however, were not able to reach consensus on a resolution pertaining to traditional forest-related knowledge, notably because of different views on indigenous rights. Neither did they pass a resolution on enhanced cooperation because they considered that UNFF had provided sufficient guidance on this item in previous sessions. Notwithstanding this latter decision, delegates stressed the importance of continued cooperation between UNFF and the UNCED conventions on biological diversity, climate change and desertification. Another outcome of UNFF-4 was the increased participation of major groups in an interactive multistakeholder dialogue that addressed issues pertaining to intellectual property rights, clear land-tenure systems, and the fair and equitable sharing of benefits arising from the sustainable management and use of forests.

Several country-led intersessional activities, organized in cooperation with organizations, helped forest experts prepare for the session on topics that included the transfer of environmentally sound technologies; monitoring, assessment and reporting; decentralization in the forest sector; and forest landscape restoration.

UNFF ad hoc expert group meets in New York

Sixty-eight experts, acting in their personal capacities, met in New York from 7 to 10 September 2004 to formulate advice to give to UNFF when it considers the future international arrangement on forests at its fifth session in May 2005.

Participants generally agreed that the international arrangement on forests needed strengthening and proposed different ways to achieve this objective – ranging from making UNFF into more of a high-level policy forum that would meet less frequently to developing a framework convention with either regional or thematic protocols. Despite lack of consensus on future modalities, experts clearly stated that maintaining the status quo was not an option. They praised CPF for its accomplishments, including improved cooperation on forest issues and strong support to country implementation of sustainable forest management.

Since UNCED, the IPF/IFF/UNFF dialogue has raised awareness of the significant contributions that forests make to the health of the planet and its inhabitants. Annual sessions of UNFF have provided a forum for continued policy development and dialogue on ways to achieve sustainable management of all types of forests, with a focus on financial and technological support as well as on monitoring progress.

Based on a review of the effectiveness of the international arrangement on forests at UNFF-5, delegates will decide on how best to continue to develop solutions, generate strong political commitment to sustainable forest management and strengthen cooperation and partnerships. Participation of all countries and advice from forestry experts are critical for a meaningful decision on – and subsequent dialogue within – the future international arrangement.

The fifth session of UNFF, including a ministerial segment and a multistakeholder dialogue, will be held from 16 to 27 May 2005 in New York, United States. High-level discussions will also take place between ministers and CPF members. In addition to making recommendations to the United Nations General Assembly on future institutional arrangements for forests, the session will address enhanced cooperation and linkages with internationally agreed development goals, especially the Millennium Development Goals.

COLLABORATIVE PARTNERSHIP ON FORESTS

CPF and its 14 members (see Box on page 60) continue to enhance cooperation and coordination on forest issues to assist countries in implementing sustainable forest management and to support UNFF. Chaired by FAO and supported by the UNFF Secretariat, CPF provides expertise and information through a focal agency system that allows for the sharing of responsibilities and builds on the comparative advantages of each member.

In collaboration with a wide range of partners, CPF helps to catalyse national, regional and international action related to forests, including the mobilization of financial resources, and helps to strengthen political commitment. Members contribute to UNFF sessions and to country-led initiatives by assisting with the preparation of documents and offering technical advice on issues within their respective mandates. Many have also seconded staff to the UNFF Secretariat.

Other international processes and bodies besides UNFF – including the United Nations Commission on Sustainable Development, CBD, UNFCCC and UNCCD – are taking note of CPF achievements and its success in bringing key organizations together.

Since its establishment in 2001, CPF has undertaken a number of joint initiatives: the online database on funding sources for sustainable forest management, streamlining reporting on forests and harmonizing forest-related definitions, among others. After the International Union of Forest Research Organizations (IUFRO) joined the partnership, CPF also became involved in the Global Forest Information Service (GFIS), an Internet gateway to forest information from around the world where users can locate maps, data sets, articles, books and other material.

CPF Sourcebook

The CPF Sourcebook on Funding for Sustainable Forest Management makes information on foreign and domestic funds accessible through an online searchable database. It contains information on some 400 potential sources of funding for forest activities and on how to develop project proposals. CPF collaborates with the National Forest Programme Facility and members of the CPF Network (see page 60) to improve and disseminate the sourcebook (available at www.fao.org/forestry/CPF-sourcebook).

Streamlining forest reporting

As part of CPF efforts to streamline reporting on forests, an Internet portal was established to provide easy access to information that countries submit to key forest-related international

Members of the Collaborative Partnership on Forests

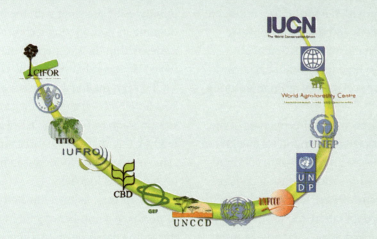

- Center for International Forestry Research (CIFOR)
- Food and Agriculture Organization of the United Nations (FAO)
- International Tropical Timber Organization (ITTO)
- International Union of Forest Research Organizations (IUFRO)
- Secretariat of the Convention on Biological Diversity (CBD)
- Secretariat of the Global Environment Facility (GEF)
- Secretariat of the United Nations Convention to Combat Desertification (UNCCD)
- Secretariat of the United Nations Forum on Forests (UNFF)
- Secretariat of the United Nations Framework Convention on Climate Change (UNFCCC)
- United Nations Development Programme (UNDP)
- United Nations Environment Programme (UNEP)
- World Agroforestry Centre (ICRAF)
- World Bank
- World Conservation Union (IUCN)

processes and organizations. Building on this initiative, FAO, ITTO, UNEP and the Secretariats of UNFF, CBD, UNCCD and UNFCCC are developing a framework for forest reporting to improve access to and coordinate information, with a view to reducing the reporting burden on countries. Searches will be structured along the seven common thematic elements of sustainable forest management (see Box on page 3). Given that consistent use of terms is a key element in any reporting process, CPF members and other partners are also working on harmonizing definitions (see www.fao.org/forestry/CPF-MAR).

CPF Network
Through an informal and open network, CPF interacts with individuals and groups that make important contributions to sustainable forest management. Thus the network enhances communication and information sharing and strengthens collaboration among organizations and processes on forest matters, for example by establishing linkages with focal points for major groups involved in UNFF deliberations. Network participants regularly receive information by e-mail about UNFF and CPF activities, and meetings are organized in conjunction with international forest gatherings.

WHAT THE POLICY DIALOGUE HAS YIELDED

Over the past 15 years there have been many positive changes in forest policy issues, although it is not possible to ascribe them all to the international dialogue on forests. They include:

- better recognition of the contributions that forests make to sustainable development;
- increased international cooperation and consensus building on complex issues;
- a greater degree of participation of civil society in decision-making;
- acknowledgement of the importance of forests to sustainable livelihoods, food security and poverty alleviation, including their relevance to the Millennium Development Goals;
- revised forest policies of multinational and bilateral donors and funding agencies;
- new financing mechanisms to capture the value of environmental services from forests, such as carbon sequestration;
- the development and implementation of national forest programmes and criteria and indicators for sustainable forest management around the world;
- new efforts to improve governance and forest law enforcement;
- the establishment of innovative partnerships at all levels.

National forest programmes have been widely embraced as a framework for developing and implementing forest policies in a participatory manner. In many countries, these programmes are proving to be effective tools for putting international forest-related commitments into practice (see Box on page 47). The use of criteria and indicators to monitor progress towards sustainable forest management is helping to improve policies, practices, information and stakeholder involvement; enhancing collaboration among countries, especially within regional processes; and contributing to the continuous improvement of FRA (see page 1). However, the degree to which countries are implementing both mechanisms varies considerably, pointing to the need for more capacity building.

CPF is recognized as a particularly useful result of the IPF/IFF/UNFF process. Increasingly, its 14 members are undertaking joint projects, cosponsoring meetings and workshops, supporting country-led initiatives and sharing information – all with a view to assisting countries to achieve sustainable forest management.

Ongoing challenges

While the IPF/IFF Proposals for Action represent a significant agreement by governments, understanding and implementing them remain a challenge. Solutions to improve financing and technology transfer continue to be elusive, with some countries and stakeholders displeased with the lack of progress. Discussions in the FAO Regional Forestry Commissions in 2002 and 2004 confirmed that implementation efforts are increasing, mostly through national forest programmes, but pointed out that the growing number of international calls for action is overwhelming implementing agencies and many developing countries. Other problems include low levels of stakeholder participation; a lack of communication between those who attend meetings and those who should implement proposals; the absence of material in national languages; and the high cost of international meetings at the expense, some would say, of providing more direct support to developing countries and countries with economies in transition.

Many countries are also concerned with the number and duplication of requests for reporting to international processes, given that many find it difficult to collect even basic information at the national level. In response, CPF members are working to streamline reporting on forests, but it will take time and resources to find practical solutions. Meanwhile, intergovernmental fora do not appear to be reducing either the number or the length of country reports they require. On the contrary, many processes seek extensive information on an annual basis through complex and overlapping questionnaires and guidelines.

The forest sector does not fare well in the face of competing demands for scarce funds, given that it contributes a relatively small share of employment and national income in most countries. While international dialogue has heightened awareness of the importance of forests for the range of benefits they provide, in most countries decision-makers have not yet taken sufficient action within their national borders – a situation that seriously hinders the implementation of sustainable forest management. In addition, other sectors often ignore the benefits of integrating forests into their policy and planning and often fail to consider forestry as an integral part of interdisciplinary natural resource management.

Some countries have expressed concern about the continued fragmentation and wasteful duplication in the work of organizations and processes despite the progress CPF is making to enhance coordination and collaboration. Processes such as CBD and UNFF attempt to

XII World Forestry Congress

In cosponsorship with FAO, the Government of Canada hosted and organized the XII World Forestry Congress in Québec City from 21 to 28 September 2003. As did previous congresses, this event served as a forum for governments, universities, civil society, the private sector and NGOs with an interest in forests to exchange views and experiences and formulate recommendations to address major forest issues at the national, regional and global levels. Some 4 000 participants from approximately 140 countries attended in their personal capacity.

Topics were considered under the theme "Forests, source of life", which was divided into three programme areas:
- forests for people – what people need from forests, ways to perceive, evaluate and use the resource, capabilities required to meet demands and the roles and responsibilities of parties concerned;
- forests for the planet – current status, trends and future outlook, and the capacity of forests to supply goods and services and to fulfil essential functions;
- people and forests in harmony – models for managing forests that take into account a number of considerations, including institutional capacity to conduct research, develop technology and enhance education.

Participants drafted a Final Statement that includes a vision, strategies and actions to achieve sustainable forest management worldwide. Recognizing that all societies depend on forests and trees for their survival and convinced that the needs of the planet and its inhabitants can be harmonized to achieve sustainable development, the congress noted the importance of building bridges with other sectors and called for continuing commitment throughout the process. Those who attended the event envisioned a future characterized by:
- social justice;
- economic benefits;
- healthy forests;
- responsible use of the resource;
- participatory, transparent and accountable governance;
- movement from dialogue to action;
- improved research, education and capacity building.

Among other prerequisites for realizing this vision, the congress identified sustained deal with the breadth of forest issues yet are often driven by one or two key concerns without giving much attention to multiple benefits, challenges, cross-sectoral dimensions or capacity building. Inconsistent messages and conflict over which body has the lead on particular issues are hindering progress in some instances.

Future of the dialogue

Over the decades, the international forest policy dialogue has contributed to bringing about many positive changes. Increasingly, however, it has become fragmented, and some processes have not lived up to the expectations of the countries that established and participate in them. With the five-year mandate of UNFF coming to a close, members will be deciding on the future international arrangement for forests in May 2005, taking into account progress that has been made, including that in other international bodies dealing with forests.

Despite developments, deforestation and

political commitment; adequate financing; close linkages with partners and other sectors; effective international cooperation; policies based on best available science and information; recognition of the culture, knowledge and good practices of indigenous peoples and local communities; and the management of forests and trees in a manner that includes interfacing with human settlements and with agroforestry, as well as other natural resource systems. Furthermore, participants agreed to promote specific strategies and actions to ensure that forests make a strong contribution to the Millennium Development Goals and other internationally agreed targets.

In adopting the Final Statement, the congress called on countries to pursue its objectives and to promote them in other sectors. It also requested that FAO publicize the statement in relevant fora and that it report to the XIII World Forestry Congress on progress in its implementation.

The XII World Forestry Congress served as a forum for governments, universities, civil society, the private sector and NGOs to exchange views and experiences

forest degradation are continuing, and illegal forest activities remain problematic. It is not enough for forest practitioners and policy-makers to reach out to other sectors to inform them of the benefits of dealing with issues in a holistic manner. Other sectors must be receptive and responsive to change. Therefore, any future international dialogue on forests should look beyond traditional forestry and establish a broader base of experts on which to draw, including those in agriculture, infrastructure development and the energy, mining and transportation sectors.

Countries should now determine why, 13 years after UNCED, worrisome trends in the forest sector are continuing, despite the many high-level commitments made. Governments and stakeholders must now ask themselves how best to support implementation efforts and bring about lasting solutions. At UNFF-5, countries must either decide to give the process a new mandate and working modalities or decide that the IPF/IFF/UNFF dialogue has yielded all it can and that it is time for other fora, instruments and processes to fill the void.

NATIONAL FOREST PROGRAMME FACILITY – A GROWING PARTNERSHIP

As reported in *State of the World's Forests 2003*, the National Forest Programme Facility is a partnership between developing countries, donors, FAO and other international organizations to stimulate the participation of stakeholders in national forest programme processes through knowledge sharing and capacity building. Hosted by FAO, the Facility operates through a multidonor trust fund under the authority of a steering committee. It began operations in 2002 and provides direct support to countries and information services worldwide. As of April 2004, it had obtained commitments from seven partners for up to US$15 million over a five-year period and was active in 36 countries: 20 in Africa, 8 in Asia and the Pacific and 8 in Latin America and the Caribbean. The Facility also supports two subregional entities in Central America.

In addition to funding workshops, training,

policy analysis and studies, the Facility supports information collection and management. It also helps to establish communities of practice where experts from institutions tackle complex issues such as stakeholder participation, funding mechanisms for sustainable forest management and payment for environmental services. Further information on communities of practices can be found at www.fao.org/forestry/site/14690/en.

After just two years of operation, the Facility had received applications from more than 60 countries and 5 subregional organizations to become partners. In addition, it is responding to an ever-growing demand for services related to national forest programmes, including the following:

- development and implementation of regional or subregional forest strategies (Central America);
- development or updating of national forest policies and programmes (Georgia, Lesotho, the Niger, Nigeria, Thailand, Tunisia);
- integration of international commitments and IPF/IFF Proposals for Action in national policy development (Lesotho, Morocco);
- integration of national forest programmes into broader national strategies (poverty reduction in Mozambique, Nigeria, Rwanda and Uganda; combating desertification in Mongolia) and intersectoral coordination in national forest programme implementation (Ecuador, Honduras, United Republic of Tanzania);
- development of subnational forest programmes (China, Senegal) and subsector strategies (South Africa);
- development and adoption of new forest legislation and dissemination of forest-related laws and regulations (Congo, Mali, Nigeria);
- development of fiscal policies (Senegal), concession systems (Mozambique), mechanisms to fund forestry (payment for environmental services in Central America), and enabling environments for private investment in the forest sector (Kenya,

The Mountain Partnership

The International Partnership for Sustainable Development in Mountain Regions (the Mountain Partnership) is a voluntary alliance dedicated to improving the lives of mountain people and protecting mountain environments around the world. Launched at the World Summit on Sustainable Development (WSSD), which was held in Johannesburg, South Africa, in 2002, the partnership taps the knowledge, expertise and vast resources of its members to support positive change in mountain areas. By mid-2004, 39 countries, 15 intergovernmental organizations and 44 groups from civil society and the private sector had joined. FAO is hosting the interim secretariat with financial support from the Governments of Italy and Switzerland and assistance from UNEP.

Building on the successes of the International Year of Mountains – 2002, enhancing the implementation of Chapter 13 of Agenda 21 and following up relevant aspects of the WSSD Plan of Implementation, the Mountain Partnership is facilitating action on the ground and working at the policy, programme and project levels. Consistent with priority concerns, members have identified initiatives pertaining to such areas as policy and law, sustainable livelihoods, watershed management, research, gender balance, education, sustainable agriculture and rural development in mountains. Activities are taking place across the Andes, Central Asia, East Africa and the Hindu Kush-Himalayas. In addition, efforts are being made to encourage linkages at the local, national, regional and global levels.

FAO/MOUNTAIN PARTNERSHIP/A. MIHICH

The Mountain Partnership promotes initiatives for sustainable livelihoods, sustainable agriculture and rural development in mountains – for example, the Andes

Malawi, Mozambique, United Republic of Tanzania);
- decentralization in the forest sector (Ecuador, Indonesia, Mali, Morocco, Mongolia, Rwanda, Uganda) and empowering local governments in forest management (Chile);
- raising stakeholder awareness of the national forest programme process (Mongolia, Namibia, Rwanda) and establishing

consultation mechanisms (Colombia, Ghana, Mali, Paraguay, South Africa, Tunisia);
- participatory forestry (Central America) and community-based forest management (Democratic Republic of the Congo, Kenya, the Philippines);
- development and application of criteria and indicators to monitor implementation of national forest programmes (Morocco);

Global Environment Facility – Operational Programme on Sustainable Land Management

GEF was established in 1991 to help developing countries and countries with economies in transition fund incremental initiatives to protect the global environment. In October 2002, the GEF Assembly added land degradation, primarily desertification and deforestation, to its other focal areas – biological diversity, international waters, climate change, ozone depletion and persistent organic pollutants. Since its inception, GEF has grown from a pilot programme to the largest single source of financing for the global environment.

The Operational Programme on Sustainable Land Management (OP#15) provides a framework for developing eligible activities to address the root causes and negative impact of land degradation on ecosystems, livelihoods and people's well-being. In the context of sustainable development, countries are expected to use participatory, integrated and cross-sectoral

approaches to deal with land degradation issues. OP#15 aims to:
• strengthen institutional and human resource capacity for planning and implementing sustainable land management;
• improve policies, regulations and incentives to facilitate wider adoption of sustainable land management practices across sectors;
• enhance the economic productivity of the land under sustainable management;
• preserve or restore the structure and functional integrity of ecosystems.

• development of forest information and monitoring systems and Web-based initiatives (China, Cuba, Honduras, Mali, Mongolia, Namibia, Paraguay, United Republic of Tanzania).

UPDATE ON FOREST-RELATED INTERNATIONAL CONVENTIONS AND AGREEMENTS
Convention on Biological Diversity
More than 2 300 participants attended the seventh Conference of the Parties (COP-7) to CBD, which was held from 9 to 20 February 2004 in Kuala Lumpur, Malaysia. Priority issues included the biological diversity of mountain ecosystems, the role of protected areas in the conservation of biological diversity, technology transfer and cooperation, and progress in achieving a significant reduction in the loss of biological diversity by 2010.

With regard to forest-related issues, delegates discussed implementation of the expanded Programme of Work on Forest Biological Diversity and requested that the Executive Secretary propose targets and develop indicators, taking into account the criteria and indicators for sustainable forest management developed by regional and international processes. The Executive Secretary was also asked to continue collaborating with other members of CPF to harmonize and streamline forest-related reporting. Decision VII/11 noted that sustainable forest management, as defined in the Forest Principles, could be considered as a means of applying the ecosystem approach to forests, and that tools developed in the context of sustainable forest management, such as criteria and indicators, national forest programmes, model forests and certification schemes, could

help to implement the ecosystem approach (see page 20).

Building on outcomes of the International Year of Mountains – 2002, COP-7 adopted a programme of work on mountain biological diversity, which makes several references to forest biological diversity. After considerable debate, COP-7 also agreed on a programme of work on protected areas that has, as one of its goals, the establishment of ecologically representative national and regional systems by 2010 for terrestrial areas and by 2012 for marine areas. Both decisions invite Parties to base implementation on national and subnational needs and to identify priorities according to their specific circumstances and conditions.

During the high-level segment, ministers reconfirmed their commitment to implement the three objectives of the convention: the conservation of biological diversity, the sustainable use of its components and the fair and equitable sharing of the benefits from the use of genetic resources. They also reiterated their pledge to work towards achieving a significant reduction in the rate of biological diversity loss by 2010.

Convention to Combat Desertification

UNCCD was adopted as a follow-up to UNCED to address threats posed by drought and desertification to the livelihoods of an estimated one billion people in more than 110 countries. The agreement came into force in 1996.

COP-6 to UNCCD took place in Havana, Cuba, from 25 August to 5 September 2003. More than 2 000 delegates participated, including some 150 NGOs and 40 international organizations from 173 countries. Among other decisions, COP-6 accepted the Global Environment Facility (GEF) as the convention's financial mechanism. Although this development is expected to further implementation efforts, much remains to be done in such areas as cooperation among developing countries and between developed and developing countries.

Delegates emphasized the convention as an important tool to eradicate poverty and called on development partners to use it in strategies to achieve the Millennium Development Goals. COP-6 also endorsed recommendations of UNCCD's Committee on Science and Technology, which encouraged institutions and NGOs to develop and test benchmarks and indicators; invited Parties to carry out pilot studies on early warning systems; and proposed the collection of case studies on traditional knowledge from local and indigenous communities.

United Nations Framework Convention on Climate Change

In December 2003, COP-9 to UNFCCC in Milan, Italy, determined that only afforestation and reforestation qualify for carbon sink projects under CDM. It also defined small-scale afforestation and reforestation projects for low-income communities and individuals. These schemes are characterized by an annual carbon sequestration of less than 8 000 tonnes of CO_2 and benefit from simpler rules and lower fixed costs. Contrary to projects in the energy sector, those in forestry may last up to 60 years. Carbon credits must either be renewed every five years or replaced when forests re-emit carbon into the atmosphere.

COP-9 also endorsed the 2003 Good Practice Guidance for Land Use, Land-Use Change and Forestry (IPCC, 2004) for assessing and reporting carbon stock changes and greenhouse gas flows in forests in the context of UNFCCC (see page 4).

Failing to reach consensus on many forestry issues in Milan, delegates resumed discussions at COP-10 to UNFCCC in Buenos Aires, Argentina, in December 2004. Issues they addressed included small-scale afforestation and reforestation projects and the use of GPG for reporting supplementary information under the Kyoto Protocol. COP-10 also addressed carbon in harvested wood products; definitions and methodologies to account for forest degradation; and methods for factoring out direct human-induced changes on forest carbon stocks from indirect and natural effects.

Developments related to climate change

By 2005, average global concentrations of carbon dioxide (CO_2), the main greenhouse gas, will have reached 380 parts per million – an increase of 36 percent since industrialization and an accumulation of 25 percent in excess of the peak concentration during the past 400 000 years (UNEP/GRID-Arendal, 2000).

In Europe, the summer of 2003 was warmer than any in the past five centuries, and precipitation has decreased 2 to 5 percent since 1900. Rainfall in the same period has decreased by between 5 and 20 percent in the Mediterranean region and northern Africa, foreshadowing the risks associated with climate change, particularly for developing countries of those regions (Bernes, 2003).

Global emissions currently amount to approximately 26.5 billion tonnes of CO_2 per year (UNEP/GRID-Arendal, 2004). Aggregate emissions of all greenhouse gases since 1990 in all industrial countries have declined by 6.6 percent, masking an actual increase of 7.5 percent in developed nations other than those with economies in transition, where emissions declined by 40 percent as a result of the collapse of many industries (UNFCCC, 2002).

Convention on International Trade in Endangered Species of Wild Fauna and Flora

International trade in wildlife involves more than 350 million plant and animal species and is estimated to be worth billions of dollars annually. The Convention on International Trade in Endangered Species of Wild Fauna and Flora (CITES) was adopted in 1973 to address unsustainable international trade of wild animal and plant species, of which more than 33 000 are listed in the treaty's three appendixes.

At COP-13 to CITES, held in Bangkok, Thailand, in October 2004, governments discussed proposals to amend Appendix II (the list of species at risk whose import and export are controlled through a permit system) and Appendix I (the list of endangered species whose commercial trade is forbidden). Plants discussed included Asia's agarwood trees (*Aquilaria* spp.), which contain valuable oil for making incense, perfumes and medicines; ramin (*Gonystylus* spp.), one of Southeast Asia's major export timbers; and yew trees (*Taxus* spp.), whose leaves are used to produce paclitaxel – a key ingredient in one of the biggest-selling cancer drugs.

Ramsar Convention on Wetlands

The Ramsar Convention on Wetlands, signed in Ramsar, Islamic Republic of Iran, in 1971, is an intergovernmental treaty that provides the framework for national action and international cooperation for the conservation and wise use of wetlands and their resources. Unlike most environmental treaties, it is not part of the UN system, although it collaborates extensively with other secretariats and partners.

As of August 2004, 1 374 sites totalling more than 121.4 million hectares were designated for inclusion in the List of Wetlands of International Importance. Since mangrove forests are underrepresented on this list, COP-8 to the Ramsar Convention in November 2002 (Valencia, Spain) adopted three resolutions emphasizing their ecological and socio-economic importance.

COP-9 to the convention will take place in Kampala, Uganda, in November 2005 under the theme "Wetlands and water: supporting life, sustaining livelihoods". Among other technical items, the management of wetlands to alleviate poverty and promote human well-being will be discussed. ◆

REFERENCES

Bernes, C. 2003. *A warmer world*. Monitor 18. Stockholm, Swedish Environmental Protection Agency.

IPCC. 2004. *Good Practice Guidance for Land Use, Land-Use Change and Forestry*. Geneva, Switzerland, Intergovernmental Panel on Climate Change (available at www.ipcc-nggip.iges.or.jp/public/gpglulucf/gpglulucf.htm).

UNEP/GRID-Arendal. 2000. *Vital climate graphics – Introduction to climate change*. Arendal, Norway, United Nations Environment Programme *Global Resources Information Database* (available at www.grida.no/climate/vital/02.htm).

UNEP/GRID-Arendal. 2004. *Greenhouse gas emissions from Annex I countries*. Arendal, Norway, UNEP/GRID (available at www.grida.no/db/maps/collection/climate9/index.cfm).

UNFCCC. 2002. *Greenhouse gas inventory database*. Bonn, Germany, United Nations Framework Convention on Climate Change (available at ghg.unfccc.int). ◆

PART II

**SELECTED CURRENT
ISSUES IN THE FOREST
SECTOR**

Enhancing economic benefits from forests: changing opportunities and challenges

Awareness of the economic, social, cultural and environmental contributions of forests and forestry has risen considerably in recent years, yet low investment and low incomes continue to plague the sector. Given its relatively small share of employment and national income – usually measured in terms of GDP – decision-makers give forestry a low priority in the face of competing demands for limited budgets. In response, attempts are being made to assess the value of all products and services, especially those pertaining to the environment. Efforts are also being undertaken to develop innovative financing mechanisms and to create markets for services in order to enhance income and to encourage investment in sustainable forest management.

Moving up the value chain and diversifying the product mix have led to a significant expansion of goods and services derived from forests. The growth of retail networks has made wood and wood products more accessible to consumers, enhancing opportunities for local communities, farmers and other resource owners in most countries. Yet the economic viability of forestry remains a concern as the sector grapples with two important issues: how to increase the size of the economic pie and how to divide it among the different segments of society.

This chapter of *State of the World's Forests 2005* analyses the contribution of the forest sector to income and describes the experiences of communities, governments and the private sector in increasing economic benefits from forests. It also identifies issues that the profession must address to make sustainable forest management an economically viable option.

The forest sector is defined in this chapter on the basis of the ILO's International Standard Industrial Classification of all Economic Activities (ISIC) (United Nations *et al.*, 2003). It includes forestry, logging and related service activities, wood industries, manufacture of wood and products of wood and cork (except furniture) and pulp and paper industries. Forestry includes the production of standing timber as well as the extraction and gathering of wild-growing forest materials except for mushrooms, truffles, berries and nuts. Forestry also includes products that undergo a minimal amount of processing, such as wood for fuel or industrial use.

FORESTS AND FORESTRY IN NATIONAL ECONOMIES
Income from forests and forest industry
Although current systems of national income accounting have serious limitations, GDP still forms the basis for assessing economic performance and allocating public funds to different sectors. Key trends related to the share of the forest sector in national income can be summarized as follows.

- Globally, the gross value added by the sector in 2000 (including forestry, logging and related activities, the manufacturing of wood, wood products, paper and paper products) is estimated at about US$354 billion, or about 1.2 percent of GDP (FAO, in preparation).
- Between 1990 and 2000, the gross value added by the sector registered a modest

growth of about 1.4 percent, while the global economy grew by about 30 percent because of gains in other sectors, especially manufacturing and services. As a result, the share of the forest sector in GDP declined from about 1.6 to 1.2 percent.

- Within the sector, the contribution of forestry *per se* remains low and appears to be declining. Globally, it accounts for about US$78 billion of the gross value added, or about 22 percent of the forest sector's contribution. Wood industries and pulp and paper make up the balance (Table 7).

The overall decline in the importance of the forest sector is consistent with that of most primary sectors, especially agriculture. The latter, which covers about 38 percent of land area and employs 44 percent of the economically active population, accounts for only about 6.2 percent of the global gross value added – ranging from 2.6 percent in developed countries to 11.9 percent in developing countries. In almost all countries, agriculture's share of GDP has declined over time (FAO, 2004a).

Interregional and intercountry differences
Considerable differences exist among regions and countries in the share of the forest sector's value added and in the contribution of subsectors (Figure 6). For example, North and Central America (mainly Canada and the United States) account for almost 40 percent of the global share of gross value added, compared

with Africa's portion of about 2 percent. The forest sector's share in the gross value added is 14 percent in North and Central America and 58 percent in Africa, while that from wood industries and the production of pulp and paper is 86 and 42 percent, respectively. Since South Africa accounts for about 42 percent of Africa's share of value added in wood industries, including pulp and paper, the rest of Africa's share in processing is much lower.

Data in this chapter are based on a number of assumptions and should therefore be interpreted with caution. However, the following observations can be made.

- The existence of large forest areas is neither an essential nor a sufficient condition for developing a vibrant sector. Indeed, many countries with low forest cover have forest industries that compete in global markets, and most of the sector's gross value added comes from wood processing rather than wood production.
- Above all, a favourable investment climate is needed to build processing capacity. The ability to develop new products and processes, knowledge of markets and entrepreneurial skills are important factors as well.

Underestimation of forestry's contribution
In the face of competing demands, foresters have only experienced moderate success in convincing decision-makers, especially in the

TABLE 7

Gross value added by the forest sector in 2000 *(million US$)*

Region	Forestry	Wood industries	Pulp and paper	Total	Contribution to GDP (%)
Africa	4 425	1 379	1 863	7 667	1.5
Asia	24 390	17 315	43 453	85 158	1.1
Europe	14 457	30 222	45 111	89 790	1.2
North and Central America	19 171	49 782	71 256	140 209	1.3
Oceania	1 176	2 553	1 655	5 384	1.3
South America	13 156	3 328	9 304	25 788	2.1
World	**76 775**	**104 579**	**172 642**	**353 996**	**1.2**

Source: FAO, in preparation.

FIGURE 6
Share of the forest sector's value added, by region and subsector

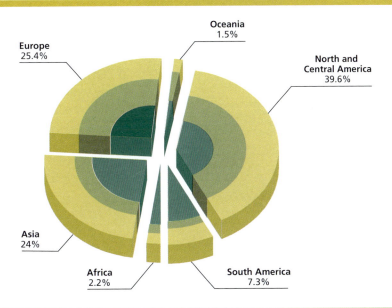

Europe
25.4%

Oceania
1.5%

North and
Central America
39.6%

Asia
24%

Africa
2.2%

South America
7.3%

Share by subsector (%)						
Subsector	Africa	Asia	Europe	Oceania	North and Central America	South America
Forests	58	29	16	22	14	51
Wood industry	18	20	34	47	35	13
Pulp and paper	24	51	50	31	51	36

Source: FAO, in preparation.

ministries of planning and finance, to allocate more resources to the sector. While political considerations guide most such decisions, neglect of the sector in national budgets is sometimes rationalized on the basis of its low contribution to income and employment, raising questions about the reliability of the system of national income accounts. Shortcomings include:

- incorrect classification of activities so that income and employment from forestry are recorded elsewhere;
- exclusion of the informal sector, which contributes significantly to income and employment in many countries;
- failure to take environmental services into account that are often critical to the performance of other sectors (watershed protection and conservation of biological diversity, for example).

The first of these shortcomings can be addressed with relative ease through improved standardization and harmonized definitions. However, lack of data makes it difficult to account for the importance of the informal sector and the value of subsistence consumption in national income statistics (Lange, 2004). Most countries have neither the resources nor the capacity to assess accurately the role of the informal sector in the overall economy.

The System of Integrated Environmental and Economic Accounting (SEEA) (see United Nations *et al.*, 2003) aims to address problems of current approaches to national income accounting. Associated satellite accounts capture changes in the flow of environmental goods and services and asset depletion. This method represents an improvement over others, but its adoption has been slow, partly owing to lack of data.

Forestry activities in the informal sector contribute significantly to income and employment in many countries – but because they are excluded from national income accounts, the economic contribution of forestry is often underestimated

FAO/R. FAIDUTTI

FACTORS THAT AFFECT VALUE CAPTURE

If at the aggregate level the forest sector is not a major contributor to national income, resource owners can still perceive it to be economically important. Their view of its profitability is influenced by the complex interaction among resource characteristics, ownership characteristics – especially socio-economic status and entrepreneurial skills – and market characteristics under different policy and institutional settings.

Resource characteristics

Depending on the characteristics of land and vegetation, including productivity, biological diversity and topography, the potential to realize economic benefits differs with the mix of products and services offered. For example, some species-rich tropical rain forests may be commercially less valuable, while the biodiversity and environmental services they provide could be significant. Similarly, vast tracts of woodlands in the dry tropics are important to local communities despite low levels of wood production and economic returns. Their value is seldom fully reflected in income statistics.

Realizing economic benefits from forests often depends on their accessibility and proximity to markets. In the 1970s and 1980s,

several commercial-scale planted forests were established without considering their viability, in particular end uses and markets. Even today, these resources often remain underutilized and poorly managed. However, large isolated forest areas offer new opportunities to provide global public goods, such as carbon sequestration and the conservation of biological diversity. Remote forests also serve as a major attraction for nature-based tourism that caters to high-value niche markets (see page 27).

Ownership characteristics

Forest ownership is in a state of flux in response to policy and legislative changes (Figure 7), reflecting society's preoccupation with balancing efficiency and equity. Although most forests are publicly owned (White and Martin, 2002), the trend towards community and private sector management is increasing, based on a common view that community and public ownership enhances social and environmental benefits, while private ownership improves economic efficiency. Weak policies and legislation encourage overexploitation and inhibit long-term investment. The following factors affect the capture of forest values.

Social and economic condition of owners. Individuals and governments with few alternative sources of income are less willing and

FIGURE 7
Forest resource ownership changes

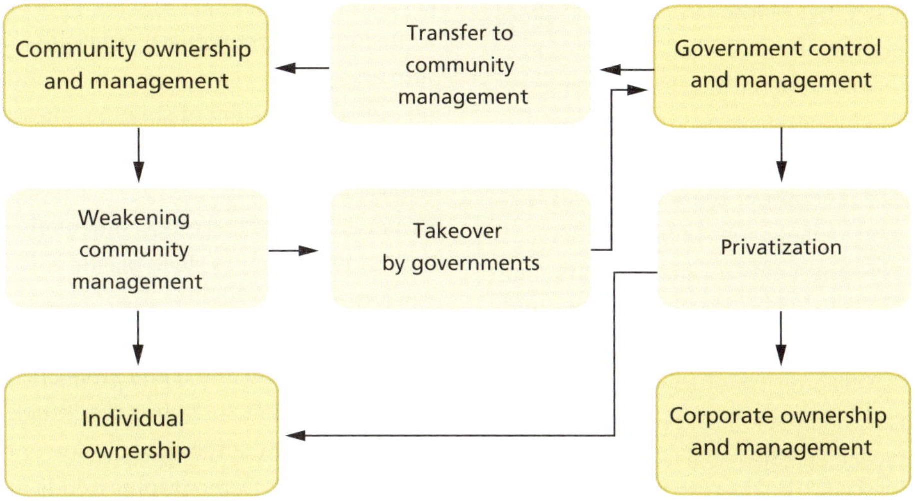

able to invest in sustainable forest management. Instead, they give a high priority to activities that require low investment and generate high economic returns over the short term. Such behaviour is also common in parts of the corporate sector, especially transnational logging companies. Governments that use revenue from forests to develop other sectors can raise funds, for example, either by selling forest products or by converting forest land to more productive uses such as cattle ranches and cash crop plantations, depending on markets. Social and economic conditions also influence access to technology and capital.

Institutional capacity. The ability to realize economic benefits from forests is tied to institutional capacity, especially the capacity to understand changing environments and the capacity to seize the opportunities they bring. Many community groups and small-scale forest owners are at a disadvantage in this regard, although the establishment of cooperatives and associations is helping to overcome some constraints. Institutional weaknesses are also evident in government management and contribute to significant leakage of benefits, including through illegal logging (see Box on

page 76). Forestry administrations in many countries are understaffed and underpaid and lack the motivation to tap the full potential of the resource. On the other hand, many corporations are able to influence markets, foresee emerging opportunities and develop strategies for the deployment of resources.

Ability to move up the value chain. Wood industries, including pulp and paper, account for a major share of the gross value added (Table 7), suggesting that moving up the value chain is key to enhancing economic benefits. The ability to do so, however, differs among owners. Profit-driven corporate ownership and management can often both identify the need for new products and services and develop them more effectively than governments and other owners. Integrating all aspects of production – from making the raw material to manufacturing the final product – has been an important strategy to increase profitability. However, many producers of wood and non-wood forest products are not in a position to set the prices, and their income is often determined by others. In the context of declining prices, sustaining primary production, including wood, often depends on direct and indirect subsidies.

Loss of income due to illegal logging

According to the World Bank, illegal logging results in a loss of US$5 billion annually and a further loss of US$10 billion to the economies of timber-producing countries. In many cases, the proportion of illegally produced timber far exceeds legal production. The activity depresses prices, undermines profitability of legitimate enterprises and helps to finance wars and civil strife. Several initiatives are addressing the problem of illegal logging, including the EU Action Plan on Forest Law Enforcement, Governance and Trade, the World Bank's Africa Forest Law Enforcement and Governance and the United States' President's Initiative Against Illegal Logging.

Market characteristics and changes

Recent decades have witnessed significant changes in the markets for forest products and services. Indications are that these will accelerate in response to changes in demography, economic performance, technology and social, political and institutional environments. At issue is the ability of resource owners to seize emerging opportunities.

Ability of consumers to pay. Markets for forest products and services are highly segmented and cater to consumers with varying ability to pay. For example, woodfuel (charcoal and fuelwood) consumed by low-income households seldom generates returns that encourage investment in production. Higher returns require the production of goods and services for high-income markets. However, this option is unavailable to many producers because of the need for large initial investments. Income from low-value products could be enhanced by increasing the quantity, but this possibility is also beyond the capacity of many small-scale entrepreneurs.

Competition. As more producers enter markets for forest products, competition is intensifying. Although the forest industry is still fragmented, some consolidation through mergers and acquisitions is taking place, especially in the pulp and paper sector. Again, such options fall outside the realm of small businesses. Market competition in highly processed items is particularly intense, exacerbated by a greater supply of less environmentally friendly commodities such as steel, plastic and concrete.

Demand for wood and wood products. The demand for wood, including woodfuel, is expected to grow, although at a slower rate than in the past. Demographic trends in many developed countries suggest a decline in demand that will drop further as recycling and processing technologies improve. On the other hand, the low per capita consumption in many developing countries suggests significant increases in demand in response to rising incomes. This trend is already evident in emerging economies such as China and India where imports of wood and wood products are escalating, bringing about important changes in the direction of global trade in forest products.

Changes in the product mix. Significant diversification of product mix has taken place in recent years with products such as medium-density fibreboard, oriented strandboard and other engineered wood products entering the market. These products often replace sawnwood, affecting the demand for large logs and thus the income of forest owners. Investments in industrial research and development are expected to accelerate the process. Efforts to cater to niche markets by customizing both wood and non-wood forest products have also been noteworthy. The rapid growth in the market for herbal products, for example, is providing new opportunities. The production and trade of secondary wood products, especially furniture and joinery, have also increased dramatically in recent years.

FIGURE 8
Recent trends in global forest product export prices

Average global export price index (1990 = 100)

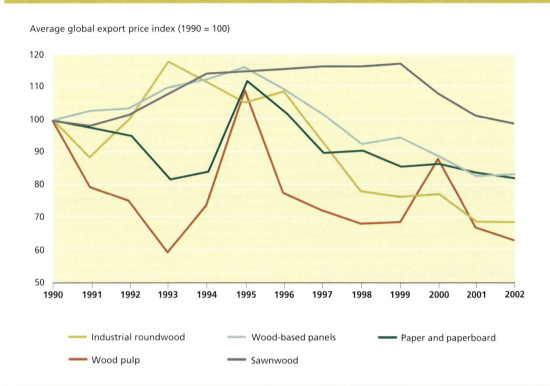

Source: FAO, 2004b.

Declining prices. Global prices for forest products have declined in the past decade (Figure 8), affecting the economic viability of the sector in many countries (New Zealand Forest Industries, 2004). In the United Kingdom, for example, the decline in stumpage fees has been significant (see Box on page 78, top) (Forestry Commission, 2002, 2004). Although deforestation in the tropics remains a concern, wood supply is not a critical problem, except in countries where limited local supply combined with a surplus capacity in wood processing have inflated prices and promoted illegal logging. In many temperate and boreal countries, removals are far below the annual allowable cut. Additional supplies from planted forests and improvement in processing technologies have also led to declining prices. On the demand side, environmental policies in developed countries have encouraged greater utilization of wood residues and recycled wood products.

In addition, concerns related to sustainability and illegal sourcing are discouraging consumers from using forest products, particularly tropical sawnwood and panel products.

Trade liberalization. Aided by improved transportation technology and trade liberalization, markets are spreading from the local to national and global levels. Several locally used products such as medicinal plants, bushmeat and ethnic foods are now exported legally and illegally, often in response to demand from people who have emigrated. Imports of low-cost forest products are increasing competition in local markets, undermining the economic viability of local production.

Markets for certified products. Markets for products certified according to specific environmental, social and economic standards have recently emerged. Consumer groups and

Income from forestry in the United Kingdom

According to United Kingdom indicators of sustainable forestry, the nominal three-year annualized return for Sitka spruce plantations declined from almost 10 percent in the period 1993–1996 to -5.4 percent in 1998–2001. This was almost entirely a result of timber prices' falling by more than 50 percent. The sector's gross value added declined from £344 million (about US$540 million) in 1995 to £298 million (about US$450 million) in 2000, or 0.04 percent of the gross value added in the economy in 2000. Most value addition is in wood processing (manufacture of wood, wood products, pulp, paper and paper products), which accounted for a gross value addition of £6 379 million (almost US$9 700 million), or about 0.64 percent of the total. Forestry, logging and related services thus accounted for less than 4 percent of the sector's contribution to gross value added, while nearly 96 percent came from processing.

Source: Forestry Commission, 2002.

Carbon markets

A recent assessment indicates a rapid growth in carbon markets, especially for project-based transactions. The market has been growing steadily from about 13 million tonnes CO_2 equivalent in 2001 to about 29 million tonnes CO_2 equivalent in 2002 and more than 70 million during the first three quarters of 2003. While developed countries are the main buyers, the share of emission reductions contracted in transition economies and in developing countries rose from 38 percent in 2001 to 60 percent in 2002 and to 91 percent during the first three quarters of 2003. Much of this increase comes from Asia and Latin America. This expansion is expected to accelerate after the Kyoto Protocol enters into force.

Source: Lecocq and Capoor, 2003.

NGOs have helped to segment these markets based on whether products are sourced from sustainably managed areas or not. Current shortages of certified products provide some limited price advantages, but this is unlikely to last as certification schemes spread. Given the high cost of implementing them, especially for small-scale producers, attempts are being made to explore alternatives such as group certification. Interestingly, most certified forests are found in boreal and temperate regions, even though certification was initiated to improve the management of tropical forests (Richards, 2004).

Markets for environmental services. The market for environmental services from forests is growing rapidly, often facilitated by national and regional policies as well as international conventions and agreements (Scherr, White and Khare, 2003). Certain segments of society that are able and willing to pay for these services are creating new opportunities for resource owners. For example, payment to protect watersheds is expected to become more widespread, especially when the linkage between upstream owners and downstream users can be institutionalized.

Market and regulatory frameworks are also being developed to address biodiversity conservation and carbon sequestration (see Box on facing page, bottom). Whether the market for environmental services will grow significantly and the extent to which it will benefit resource owners remain uncertain (Landell-Mills and Porras, 2002). For example, the substantial economic benefits anticipated from biodiversity prospecting ten years ago have not yet been realized (Katila and Puustjärvi, 2003; Laird and ten Kate, 2002).

ECONOMIC BENEFITS TO FOREST RESOURCE MANAGERS
Local communities
Increased recognition of the role of communities in protecting and managing forests in the past two decades has led to a major shift in forestry development (Alden Wily, 2003). Joint forest management and forest user groups have increased community participation and helped to achieve economic, social and environmental goals that governments sometimes have difficulty meeting. Although much remains to be done, in many countries the rights of indigenous communities to own, use and manage forests and other natural resources are being recognized. For communities to take advantage of emerging opportunities, the following are needed:

- policy and legal frameworks that protect community rights over resources;
- access and proximity to markets;
- expertise and access to information, especially on markets and prices;
- institutional capacity to manage resources, add value to products and services and negotiate with other players.

Despite the benefits arising from community ownership and management, pitfalls have also been identified. Transfer of responsibilities is often limited to forests with little commercial value. Low productivity of such areas implies the need for significant investment and for effective institutions to capture and distribute benefits equitably. While communities are in a good position to identify and cater to local

> ### Income from logging to customary owners in Papua New Guinea
>
> Although communities in Papua New Guinea control 97 percent of the land, they have little say in the operations of foreign companies that are awarded logging contracts. Customary landowners receive about 12 percent of the value of logs through a complicated and indirect system of payments from central government or from local funds that are intended to facilitate development but are open to mismanagement. Current approaches largely aim to generate revenue for the government.
>
> *Source*: Hunt, 2002.

needs, they are less able to deal with national and global markets. Isolated communities face high transaction costs and have difficulty understanding consumer needs, adapting production to changing requirements and linking products with end users. Overcoming such constraints often depends on external support.

Communities that own valuable forest resources, as in Papua New Guinea (see Box above), face similar challenges to those that gain control through policy and legal changes. Interacting with external markets requires a good understanding of changing demand and prices and the ability to negotiate with logging companies and wood industries. Institutional weaknesses and lack of information undermine the capacity of communities to take full advantage of economic benefits and, consequently, they receive only a fraction of the income. The opposite holds true where institutional arrangements are well developed, communities are well informed and industries depend on local wood supplies. A recent study on revenue capture by native landowners in Fiji revealed that well-informed communities receive a significant share – about 85 percent –

FIGURE 9
Income to forest owners in Fiji *(F$/m³)*

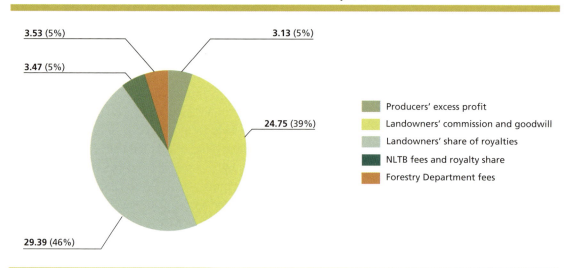

3.53 (5%)   **3.13** (5%)

3.47 (5%)

24.75 (39%)

29.39 (46%)

- Producers' excess profit
- Landowners' commission and goodwill
- Landowners' share of royalties
- NLTB fees and royalty share
- Forestry Department fees

Source: Whiteman, 2004.
Notes: 1 Fiji dollar (F$) = US$0.53 (2003).
NLTB = Native Land Trust Board.

of the value of wood obtained from land under customary ownership, through royalties fixed by the Native Land Trust Board and through direct negotiations with concessionaires (Whiteman, 2004) (Figure 9).

Where forests are more productive and valuable, the involvement of communities in their management has been limited (Oyono, 2004) and mostly at the insistence of NGOs or donor organizations. Even when communities are assigned valuable forests,

---

Criteria of fair trade certification

- Fair prices for farmers and decent working and living conditions for workers
- Direct trade with farmers, bypassing inter-mediaries
- Free association of workers and cooperatives, with structures for democratic decision-making
- Access to capital
- Sustainable agricultural practices, including restricted use of agrochemicals

---

insufficient information on markets and technology and weak institutional frameworks limit their ability to benefit. Although many countries are now involving communities in the management of wildlife, low financial returns are discouraging their participation (Akumsi, 2003).

With knowledge becoming an important source of wealth, significant efforts are being made to protect intellectual property rights and develop mechanisms for the equitable sharing of benefits arising from the use of traditional knowledge with local communities. However, the degree to which intellectual property rights protect this type of knowledge varies considerably, but partnerships with research institutions and enterprises are helping (see Box on facing page).

Fair trade certification (see Box on the left), which attests that procured goods adhere to well-defined environmental and social criteria, including payment of fair prices to producers, has been attempted with NWFPs such as babassu oil. However, this type of scheme only covers a fraction of trade in such products.

Benefiting from traditional knowledge

The forest-dwelling Kani community in the Indian State of Kerala traditionally uses the fruits and leaves of *Trichopus zeylanicus* (locally known as *arogyapacha* or health-herb) for its anti-fatigue and anti-stress properties. Based on information from the community, a public-sector organization conducted research and registered two national patents in 1996. A pharmaceutical company paid a licence fee to the organization to produce and market the drug, which the research institute and community share equally through a community-managed trust fund. The success of this arrangement has largely been a result of:

- the role of individuals in the research organization and of civil society initiatives to compensate the local community;
- effective local research and development capacity;
- linking research with production and marketing to add value;
- transparent arrangements and an effective legal framework.

Source: La Vina, 2002; UNDP, 2004.

Governments

In many countries, ownership and management of most forests are in the public domain, justified by the need to protect the wealth of the nation. Other reasons include the failure of markets to achieve social goals or to provide public goods such as watershed protection and biodiversity conservation. Inevitably, the involvement of governments entails choosing between competing objectives, sometimes at the expense of economic efficiency. Revenue from forests is often used to finance the development of other sectors rather than to reinvest in sustaining production. In several countries, governments also manage wood industries and are involved in wood processing on social grounds and to develop rural areas. However, in recent years the management of wood industries is increasingly being divested to the private sector.

State of forests and value capture. Public-sector management in many countries focuses on forests that have commercial potential or fulfil critical environmental functions. Forests considered of low value require significant investment – often beyond the means of governments – and the costs to protect such areas are much greater than the revenue they generate. Hence, they are the first to be transferred to the private sector or to communities.

Governments manage high-value forests, either directly or through concessionaires, primarily for timber production. The priority assigned to wood production has led to most other forest products being termed "minor forest products" because of the small contribution they make to government revenue. In addition, national parks and game reserves in most countries are managed for social and environmental benefits, notwithstanding the low incomes they generate. As a result of declining returns from wood production caused by falling prices and the exclusion of large tracts of forests, forestry agencies are paying more attention to service functions such as recreation and charging for them (Leslie, 2003).

Institutional arrangements for value capture. Although markets for environmental services have expanded, wood production remains the most important source of income from forests. Where forests have limited potential for revenue capture, management costs are high, resulting in their neglect and deterioration. As governments move out of wood processing and value addition, they have to pay more attention to capturing the full income from wood production and forest-derived services. Various approaches to achieve this are noted as follows.

- *Market-based price determination.* Market-based prices, primarily through tender or other bidding systems that

enhance competition, are replacing charges determined arbitrarily. Yet, in many countries, administrative approaches prevail and undermine efforts to capture the full potential income. Moreover, market imperfections persist because of monopolies or oligopolies in the production, and in some countries, in the purchase, of wood. To enhance value capture, substantial market research would be required to understand changing demands, supply and prices. However, most public-sector forestry organizations are ill equipped to conduct such studies, making price fixation susceptible to non-economic considerations, including rent-seeking by vested interests.

• *Improving tax collection.* In many countries, forestry administrations lack the capacity to collect royalties, as inadequate as they may be, and institutional competence has not kept pace with the rapid expansion in logging. Hiring independent firms to collect taxes is one way to address corruption, illegal logging and loss of income to governments. Cambodia, Cameroon, Ecuador, Papua New Guinea and Suriname have tried this approach with varying degrees of success. Inspection and tax collection at exit points

are not economically viable if products can be moved out of the country through several points. Moreover, control at exits does not ensure the sustainability of wood production.

• *Separating revenue capture from other government functions.* Most forestry organizations in the public sector find it difficult to fulfil administrative and commercial functions simultaneously, especially when objectives conflict. Assigning business-related functions to more autonomous bodies such as corporations, authorities and boards is one solution. The success of such arrangements largely depends on efficient management and the extent to which businesses can operate freely and flexibly. An effective auditing system that allows public oversight is critical to ensure economic efficiency. In addition, the long-term viability of forest enterprises will depend on their adapting to rapidly changing opportunities.

• *Privatizing commercial functions.* The establishment of quasi-state commercial enterprises has not always improved the ability of governments to capture the full income from forests. Several countries have privatized wood industries and

TABLE 8

Revenue from the management of state forests in Eastern Europe, 1999 to 2001

Country	Employees per 1 000 ha of forests	Revenue per ha (€)	Transfers to/from government per ha (€)	Funds per ha after transfer (€)
Bulgaria	2.0	12.1	+1.4	13.5
Czech Republic	2.6	330.8	+4.1	334.9
Estonia	1.7	69.1	-16.4	52.7
Hungary	10.6	185.7	-1.3	184.4
Latvia	0.4	22.0	-10.0	11.9
Lithuania	7.5	81.0	-6.2	74.8
Poland	4.9	123.4	-0.7	122.7
Romania	5.5	28.8	-1.4	27.3
Slovakia	13.4	120.3	+5.8	126.1
Slovenia	n/a	91.0	-9.1	81.9
Turkey	1.1	20.3	+9.9	30.2

Note: n/a = not available.
Source: Simula, 2003.

planted forests as part of a larger policy of economic liberalization. In many former centrally planned economies, forests are being returned to previous owners. In other instances, governments are attempting to divest themselves of money-losing enterprises, but potential buyers are primarily interested in acquiring businesses that are profitable. Managing the privatization process is not without problems, including the significant potential for vested interests to misappropriate funds; the undervaluation of assets of enterprises as a result of a lack of transparency and professional competence; and social concerns, especially regarding employment.

Economic viability of public forestry. Even if they own extensive tracts of forests, many government organizations struggle to make ends meet because net revenues are very low and because they have limited ability to capture the full economic potential from the resource. Another reason is high management costs in some countries, for example in Eastern Europe (except Estonia and Latvia) (Table 8). Studies on fiscal policies in Africa reveal a similar situation. Harvesting old-growth or mature forests generates high returns to governments, provided the institutional capacity is adequate to prevent leakages.

Small-scale owners

Policy and institutional changes are creating new opportunities for farmers and other landowners, resulting in increased investment in small-scale tree cultivation and other activities, including wood processing. In a few places, landowners are also managing private conservation areas, taking advantage of ecotourism and the benefits that this growing industry is bringing (see page 27). The decision to go this route largely depends on market opportunities; the social and economic situation, including resource ownership; and institutional capacity. In broad terms, private-sector forest management ranges from low-intensity systems, where owners increase incomes based on a number of products

> **Investments and returns to Chinese farmers from forestry**
>
> A rural household survey in China revealed that:
> - in 2001, the average household expendi- ture on forestry was 0.61 percent of the total;
> - in 1999, 2 percent of the national labour force was involved in forestry;
> - of the land and water pond areas that rural households managed, 27 percent were forest lands;
> - in 2001, income from forestry accounted for about 1.5 percent of net household income.
>
> *Source:* Zhang, 2004.

and services, to intensively managed systems that focus on one or a few items.

Low-intensity management systems. Trees form an integral part of many farming systems such as home gardens in the humid tropics and agroforestry parklands in Sahelian West Africa. Holdings are usually small, and limited local demand encourages a low-investment/low-return management regime. Often, owners maintain trees and other vegetation for the social, cultural and environmental benefits they provide rather than for economic reasons (see Box above). However, these resources are an important source of products and income in emergencies. Since shortage of labour and lack of investment funds constrain most farmers, land-use intensity is low, and reluctance to take risks is high.

Intensively managed systems. Expanding markets and declining supplies of wood and other products from natural forests are encouraging farmers to plant more trees and cultivate plants for NWFPs, including medicines. Thus, the proportion of wood from farms has risen in recent years and, in some countries, is surpassing the share originating from forests.

Farmers also plant trees to hold on to land for future security or for speculative purposes, especially if they have alternative sources of income. Increasing demand for herbal products has led to intensive cultivation of popular items, mostly for overseas markets.

Outgrower schemes and other partnership arrangements between industry and landowners are emerging as well (Mayers and Vermeulen, 2002). Industry often provides improved planting materials and technical advice on management practices and agrees to buy wood at market prices at the end of the rotation. While this approach benefits smallholders, industry also gains by reducing risks associated with owning and managing large plantations.

The rising demand for nature-based recreation has led to increased private-sector involvement in the management of parks and game reserves, for example in Costa Rica, Kenya, Namibia, South Africa and the United States. In South Africa, private protected areas exceed those that are publicly owned and managed (Katila and Puustjärvi, 2003). Game management is a low-intensity option based on ownership of extensive areas, the presence of wildlife populations and unique natural environments. Adding value involves improving access, marketing and building facilities for visitors. Several private game reserves are providing packages that cater to the different needs of customers, and many owners are forming partnerships to manage large conservation areas jointly.

Corporations

Corporations are major players in forestry, including in the management of forests, logging and wood processing, and are a driving force behind the globalization of the sector, capable of moving investment, technology and raw material transnationally. Investors fall into two categories: those focused on logging and those who integrate forest management with wood industries. Opportunities for short-term investments in logging, with little regard for sustainability, have recently expanded in a number of countries. Investors have been taking advantage of weak policies and institutions

to earn substantial income by logging above permissible quantities and outside concession areas, undermeasuring, using transfer pricing and evading taxes. These types of operators have created considerable uncertainty in the timber industry, undermining the economic viability of legitimate investments. Most corporations, however, take a long-term approach to resource management, investing in improvements as well as in processing, with a view to enhancing value capture through value addition, reducing costs through better technologies and increasing their market share as described below.

Value addition. Since most income is derived from processing and marketing, corporate efforts have focused on developing new products and services, aided by technologies that help to reduce labour costs and requirements for raw materials. Most corporate players are investing significantly in research and development of new products and processes and are using

Timberland investment management organizations

As forest industry began divesting forest ownership, groups of forest owners formed timberland investment management organizations (TIMOs) to take advantage of emerging opportunities. United States-based TIMOs expanded operations to several other countries, especially in the Southern Hemisphere – Argentina, Brazil, Chile, New Zealand and Uruguay. Low unit prices, long-term potential for value appreciation and high productivity increased institutional investment in timberland from about US$1 billion in 1989 to about US$14.4 billion in 2002, over an area of about 7.4 million hectares.

Source: Ravenel, Tyrrell and Mendelsohn, 2002.

technologies to improve trees and propagate clones in the context of planted forests.

Fibre security. A secure supply of raw material provides a competitive advantage to large wood industries, prompting a number of them to buy extensive tracts of forests or acquire large concessions, especially if the cost is low. Private corporations manage them, feeding processing units that may be spread across several countries. Moreover, investor-friendly legislation and incentives such as direct and indirect subsidies have contributed to the rapid expansion of industrial plantations (Enters, Durst and Brown, 2003). Improved management and the wider application of science and technology, including better processing technologies, have significantly boosted productivity and increased wood supplies, thereby reducing concerns. Consequently, many corporations are less inclined to hold on to their forest assets and are paying more attention to processing – their core competence.

Asset valuation and stockholder pressures. Another reason that large corporations are divesting their forestry assets is that they must now value them at market prices. Holding on to land and forests when wood prices are declining is having a negative impact on balance sheets and compelling a number of them to sell off the forestry component of their enterprises. As a result, timber management organizations are emerging to buy assets at low prices and manage them on behalf of investors who seek low-risk, but stable, long-term returns (see Box on facing page) (Neilson, 2003). A turbulent market in forest property has also provided opportunities for short-term investors to buy low and sell as soon as the market experiences an upturn.

Global expansion through new investments and mergers and acquisitions. With competition becoming more intense, adding value alone is not sufficient to ensure survival. Expanding to new markets and consolidating through mergers and acquisitions are important strategies for many corporations. In their quest to dominate

the marketplace, popular tactics include buying out and reorganizing losing firms, achieving economies of scale, reducing personnel and moving production to countries with low labour costs. In recent years, many leading transnational forestry corporations have shifted their operations outside their home country. For example, in 2003, 59 percent of the paper and paperboard capacity of Finnish companies was located outside Finland (Finnish Forest Industries Association, 2004).

CONCLUSION: VALUE, UTILITY AND ECONOMIC BENEFITS

Realizing the economic benefits from forests is complex and depends on a variety of factors. As with other primary sectors such as agriculture, forestry's share in national income is declining, along with profitability. Although environmental and other values that forests provide are gaining recognition, wood and wood products will remain an important source of income to most owners in the immediate future. Therefore, governments and other owners of the resource must endeavour to capture the full potential arising from wood production. Creating conditions for the development of efficient markets, including combating illegal logging, is a prerequisite. Moving up the value chain is another strategy, but because of various constraints it is not open to everyone. With wood supplies increasing, more needs to be done to promote wood as an environmentally friendly and energy-efficient product.

In some instances, resource owners may not be in a position to transform all current and potential uses of forests into economic benefits. Societies at various stages of development assign different values to products and services and, at any given time, direct only a small portion through the marketplace. The farmer who owns a few trees, a government forestry department that owns large tracts of forest, or a forest industry that manages a block of planted forests is not always measuring economic benefits from each investment component. Essentially, the focus is on enhancing all the benefits, only some of which are captured in monetary terms.

As society evolves, new demands arise and products and services that previously had no perceived benefit or price tag become important.

Putting a price tag on or creating markets for goods and services derived from forests is assumed to enhance investment in sustainable forest management. However, results have been mixed, as not all forest benefits can be practically exchanged in the marketplace. Notwithstanding efforts to date, a significant portion of forest goods and services will remain outside markets, preventing resource owners from appropriating any associated revenue. Therefore, society at large must cover the costs of providing such goods and services.

The justification for forests and forestry, as with other human pursuits, needs to progress beyond the narrow domain of economic benefits. This requires society to take a broader view of the sector. The forestry profession must also convince decision-makers to look past national income estimates, recognize the significance of conserving natural resources and look beyond their market prices. ◆

REFERENCES

Akumsi, A. 2003. Community participation in wildlife management: the Mount Cameroon experience. *Unasylva*, 214/215: 37–42 (also available at www.fao.org/forestry/unasylva).

Alden Wily, L. 2003. *From meeting needs to honouring rights: the evolution of community forestry.* Presented at the XII World Forestry Congress, Québec City, Canada.

Enters, T., Durst, P.B. & Brown, C. 2003. What does it take to promote forest plantation development? Incentives for tree-growing in countries of the Pacific rim. *Unasylva*, 212: 11–18 (also available at www.fao.org/forestry/unasylva).

FAO. 2004a. *The State of Food and Agriculture, 2003–04.* Rome.

FAO. 2004b. *FAOSTAT Forestry data.* Rome (available at apps.fao.org/faostat/collections?version=ext& hasbulk=0&subset=forestry).

Finnish Forest Industries Association. 2004. *Facts and figures* (available at english.forestindustries.fi/ figures).

Forestry Commission (UK). 2002. *Indicators of sustainable forestry: economic aspects* (available at www.forestry.gov.uk/forestry/INFD-4xHDBF).

Forestry Commission. 2004. *National statistics: coniferous standing sales price index, 27 May 2004.* Edinburgh, UK.

Hunt, C. 2002. *Production, privatisation and preservation in Papua New Guinea forestry.* Instruments for Sustainable Private Sector Forestry series. London, International Institute for Environment and Development.

Katila, M. & Puustjärvi, E. 2003. *Impact of new markets for environment services on forest products trade.* Rome, FAO. (Unpublished)

Laird, S.A. & ten Kate, K. 2002. Linking biodiversity prospecting and forest conservation. *In* S. Pagiola, J. Bishop & N. Landell-Mills, eds. *Selling forest environmental services. Market-based mechanisms for conservation and development.* London, Earthscan.

Landell-Mills, N. & Porras, I.T. 2002. *Silver bullet or fools' gold: a global review of markets for forest environmental services and their impact on the poor.* London, International Institute for Environment and Development.

Lange, G.-M. 2004. *Manual for environmental and economic accounts for forestry: a tool for cross-sectoral policy analysis.* Working Paper, Forestry Department. Rome, FAO (available at www.fao.org/documents/ show_cdr.asp?url_file=/docrep/007/j1972e/ j1972e00.htm).

La Vina, A.G.M. 2002. *The emerging global regime on genetic resources: its implications for local communities.* Working Paper: Globalization, Environment and Communities. Washington, DC, World Resources Institute.

Lecocq, F. & Capoor, K. 2003. *State and trends in carbon market 2003.* World Bank Carbon Finance Business Team. Washington, DC, World Bank.

Leslie, R. 2003. Charging for forest recreation. *Unasylva*, 212: 25–30.

Mayers, S. & Vermeulen, S. 2002. *Company-community partnerships: from raw deals to mutual gains?* Instruments for Sustainable Private Sector Forestry series. London, International Institute for Environment and Development.

Neilson, D.A. 2003. Forest ownership by corporates – a thing of the past? *New Zealand Journal of Forestry*, 48(1): 3–8.

New Zealand Forest Industries. 2004. *Market notes.* New Zealand Forest Industries, June 2004 (available at www.nzforest.com).

Oyono, P.R. 2004. One step forward, two steps backward? Paradoxes of natural resources management decentralisation in Cameroon. *Journal of Modern African Studies*, 42(1): 91–111.

Ravenel, R., Tyrrell, M. & Mendelsohn, R. 2002. *Institutional timberland investment: a summary of a forum exploring changing ownership patterns and the implications for conservation of environmental values.* Yale Forest Forum Series, 5(3). New Haven, USA, School of Forestry and Environmental Studies, Yale University.

Richards, M. 2004. *Certification in complex socio-political settings: looking forward to the next decade.* Washington, DC, Forest Trends.

Scherr, S., White, A. & Khare, A. 2003. *Current status and future potential markets for ecosystem services of tropical forests: an overview.* Report prepared for the International Tropical Timber Organization. Washington, DC, Forest Trends.

Simula, M. 2003. Forest sector reforms in Eastern European countries – overview and lessons learnt.

In *Institutional changes in forest management in countries with transition economies: problems and solutions: Workshop Proceedings*, 25 February 2003, Moscow.

UNDP. 2004. *Equator prize 2002: finalists and winners.* Kerala Kani Samudaya Kshema Trust, United Nations Development Programme (available at www.undp.org/equatorinitiative/EquatorNet/indiaPage.htm).

United Nations, European Commission, International Monetary Fund, Organisation for Economic Co-operation and Development & World Bank. 2003. *Integrated environmental and economic accounting.* New York, USA, United Nations.

White, A. & Martin, A. 2002. *Who owns the world's forests? Forest tenure and public forests in transition.* Washington, DC, Forest Trends.

Whiteman, A. 2004. *A review of the forest revenue system and taxation of the forestry sector in Fiji.* Draft report for the Fiji Ministry of Fisheries and Forests and FAO. Rome.

Zhang, K. 2004. How much the forests mean to farmers in China. *APANews*, 23: 6–7. ◆

Realizing the economic benefits of agroforestry: experiences, lessons and challenges

Agroforestry is the set of land-use practices involving the deliberate combination of trees, agricultural crops and/or animals on the same land management unit in some form of spatial arrangement or temporal sequence (Lundgren and Raintree, 1982). Cultivating trees in combination with crops and livestock is an ancient practice. However, several factors have contributed to a rising interest in agroforestry since the 1970s: the deteriorating economic situation in many parts of the developing world; increased tropical deforestation; degradation and scarcity of land because of population pressures; and growing interest in farming systems, intercropping and the environment (Nair, 1993). Most research on agroforestry has been conducted from the biophysical perspective, but socio-economic aspects are gaining attention (Mercer and Miller, 1998).

Main agroforestry practices include improved fallows, taungya (growing annual agricultural crops during the establishment of a forestry plantation), home gardens, alley cropping, growing multipurpose trees and shrubs on farmland, boundary planting, farm woodlots, orchards or tree gardens, plantation/crop combinations, shelterbelts, windbreaks, conservation hedges, fodder banks, live fences, trees on pasture and apiculture with trees (Nair, 1993; Sinclair, 1999).

EXAMPLES OF ECONOMIC BENEFITS OF AGROFORESTRY PRACTICES

Agroforestry practices differ considerably from country to country as farmers adapt to needs and circumstances. This section provides a number of examples of the agroforestry strategies successfully employed by farmers in different situations.

Fodder

Farmers and pastoralists have long used fodder trees and shrubs to feed their livestock, but traditional practices tend to be extensive, with farmers lopping off branches or allowing their animals to browse. Integrating trees into systems where they can be planted close to each other and pruned or browsed intensively can help increase economic benefits.

In the highlands of central Kenya, for example, farmers plant fodder shrubs, especially *Calliandra calothyrsus* and *Leucaena trichandra*, to use as feed for their stall-fed dairy cows (Franzel, Wambugu and Tuwei, 2003). The farm-grown fodder increases milk production and can substitute for relatively expensive purchased dairy meal, thus increasing farmers' income. Fodder shrubs also conserve the soil, supply fuelwood and provide bee forage for honey production. Rather than cash outlays, farmers only need small amounts of land and labour to plant them. Some farmers also earn money by selling seeds.

In Cagayan de Oro, Philippines, a combination of improved fodder grasses and trees (*Gliricidia sepium*) has helped farmers increase income from livestock production, increase crop production and reduce farm labour, especially for herding and tethering (Bosma *et al.*, 2003).

Agroforestry systems for fodder are also profitable in developed countries. In the northern agricultural region of western Australia, tagasaste (*Chamaecytisus proliferus*) planted in alley farming and plantation systems has increased returns to

FAO/FO-0052/F. OHLER

*Cultivating trees in combination
with crops and livestock is an
ancient practice, but interest in
agroforestry has been rising since
the 1970s, with socio-economic
aspects now gaining attention*

farmers whose cattle formerly grazed on annual
grasses and legumes (Abadi *et al.*, 2003).

Soil fertility

With intensified agriculture and reduced
fallowing periods, soil fertility has emerged
as a key problem in many farming systems
throughout the tropics. In several areas,
researchers and farmers have developed
improved tree fallows as one means to increase
crop yields.

In Malawi and Zambia, for example, planting
the shrubs *Tephrosia vogelii*, *Sesbania sesban*,
Gliricidia sepium or *Cajanus cajan* in fallows for
two years, cutting them back, then following
them with two to three years of maize
cultivation increased maize yields compared
with planting continuous unfertilized maize
(Franzel, Phiri and Kwesiga, 2002). Although
fertilized maize was found to perform even
better than improved fallows, the fallows
strategy proved beneficial to farmers who could
not afford fertilizer.

Another agroforestry practice for improving
soil fertility is biomass transfer – the manual
transfer of green manure to crops – which
increases vegetable yields, extends the
harvesting season and improves the quality of
produce. In western Kenya, farmers who treated
their vegetable plots with leaves from *Tithonia*

diversifolia hedges grown along field boundaries,
together with small amounts of phosphorus
fertilizer, doubled their returns to labour
(Place *et al.*, 2002).

Timber and fuelwood

Agroforestry produces timber and fuelwood
throughout the world. For example,
intercropping of trees and crops is practised
on 3 million hectares in China (Sen, 1991).
Farmers intercrop *Paulownia* spp. (primarily
P. elongata) with cereals over a wide expanse of
the North China Plain. The tree is deep rooted,
interferes little with crops and produces high-
quality timber (Wu and Zhu, 1997). In Minquan
County (Henan Province), 30 years after the
introduction of agroforestry, two-thirds of the
46 000 ha of farmland were intercropped with
trees of this genus. In one commune, *Paulownia*
spp. accounted for 37 percent of farm income
(Wu and Zhu, 1997). In addition to timber, these
species provide excellent fuelwood, leaves for
fodder and compost fertilizer and protection
against wind erosion and evapotranspiration
(Wu and Zhu, 1997).

In Tabora District, United Republic of
Tanzania, about 1 000 tobacco farmers have
started *Acacia crassicarpa* woodlots to produce
fuelwood for tobacco curing, intercropping
the trees with maize during the first two years
(Ramadhani, Otsyina and Franzel, 2002).
Growing wood on farms prevents the felling of
trees from the forest, reducing forest degradation
and saving costs of transporting fuelwood.

In Uttar Pradesh, India, 30 000 farmers grow
poplar (*Populus deltoides*) to sell to the match
industry on woodlots that average 1.3 ha.
Intercropping is common, especially in the
first two to three years (Jain and Singh, 2000;
Scherr, 2004).

In the United Kingdom, a range of timber/
cereal and timber/pasture systems has been
profitable to farmers. McAdam, Thomas and
Willis (1999) found that ash trees intercropped
with ryegrass pastures did not influence the
pasture yields for the first 10 years of the
40-year rotation. Incentives to increase
biodiversity in pastoral systems and the

Gum arabic husbandry

Through centuries of practice, gum producers in sub-Saharan Africa have devised a comprehensive protocol, from tree management to tapping, collecting, cleaning, sorting and marketing. Over the years they have learned that gum trees (*Acacia senegal*) are ripe for tapping after a dormant period following the rainy season and judge the best time for this activity by the shedding of leaves, a change in the colour of bark and, for experienced elders, by the smell of stripped bark. The first gum exudation takes place a few weeks after tapping and then is harvested in a series of pickings.

More than just providing a commercial product, gum trees supply a number of goods and services to farmers. Because of its deep tap roots and wide lateral root system – up to 40 percent of biomass may be underground – the tree is

highly valued as a soil stabilizer. In sandy areas, it assists in dune fixation, acts as a buffer against wind erosion and decreases water runoff. Its local value derives in part from the belief that, in traditional rotations, crops have higher yields after *A. senegal* fallow. The tree is also a source of fodder and browse, as well as fuelwood.

As a well-established activity, gum production has all the ingredients for growth and sustainability in place, including policies, legislation and institutional capacity for resource management, development and quality control (Chikamai, 1996).

uncertainty of meat prices versus timber prices further encourage farmers to practise agroforestry.

Environmental services: windbreaks, carbon sequestration and biodiversity

Studies of the environmental benefits of agroforestry are far fewer than those related to economic benefits, and studies seeking to monetize such benefits are almost non-existent. Available information indicates that agroforestry can provide a greater range of environmental benefits than conventional types of annual crop cultivation. For example, Murniati, Garrity and Gintings (2001) found that in areas adjacent to national parks in Sumatra, Indonesia, households with diversified farming systems, including mixed perennial gardens, depended much less on gathering forest products than did farms cultivating only wetland rice. Thus, tree felling and unsustainable hunting practices in the nearby parks were reduced. The findings suggest that promoting diversified farms with agroforestry in buffer zones can enhance forest integrity.

Windbreaks are one of the oldest agroforestry

systems in North America. In the Canadian prairies, more than 43 000 km of windbreaks have been planted since 1937, protecting 700 000 ha. In 1987, approximately 858 000 windbreaks in the United States, mostly in the north central and Great Plains areas, spanned 281 000 km and protected 546 000 ha (Williams *et al.*, 1997). Kort (1988) estimated the yield increase of crops sheltered from wind to be 8 percent for spring wheat, 12 percent for maize, 23 percent for winter wheat and 25 percent for barley. In addition, windbreaks improve crop water use and protect livestock and homesteads.

Several examples exist of private companies supporting agroforestry in exchange for carbon benefits. In the Scolel-Té pilot project in southern Mexico, 400 small-scale farmers in 20 communities are converting from swidden agriculture to agroforestry, either by intercropping timber trees with crops or by enriching fallow lands (de Jong, Tipper and Montoya-Gomez, 2000). The International Federation of Automobiles has purchased the resulting 17 000 tonnes of carbon offsets for US$10 to $12 per tonne of carbon. Sixty percent of the revenues have gone to farmers. However,

the question remains whether returns from agroforestry will be sufficient for farmers to maintain the practices once carbon payments have ended (de Jong, Tipper and Montoya-Gomez, 2000). Similarly, in the highlands of Ecuador, farmers participating in a carbon-trading project are planting mixed woodlots of pine, eucalyptus and indigenous species. Pine and eucalyptus are profitable, but the slow-growing indigenous species offer negative returns. This again puts into question the sustainability of carbon-trading tree projects involving activities that are not in themselves profitable (Smith and Scherr, 2002).

Gockowski, Nkamleu and Wendt (2001) compared the environmental benefits of the most prevalent cropping practices around Yaoundé, Cameroon: cocoa agroforests and food crops rotated with short or long fallows. Cocoa agroforests ranked first in carbon stocks, numbers of plant species and degree of plant biodiversity. They also ranked highest in terms of social profitability – the economic returns from society's perspective, not taking into account the effects of taxes, subsidies and distorted exchange rates. However, with regard to the most important criterion to farmers, net returns to labour, there was little difference among the alternatives.

MULTIPLE STAKEHOLDERS AND MULTIPLE CRITERIA FOR ASSESSING BENEFITS

Most economic analyses of agroforestry focus on benefits to farmers, yet many groups of stakeholders are interested in changes of land use. Tomich et al. (2001) used a matrix to assess how various land-use practices performed across different criteria important to six groups in Sumatra: the international community, hunter-gatherers, small-scale farmers, large-scale estates, absentee farmers and policy-makers. The results showed that while sound management of natural forests is most conducive to achieving carbon sequestration and biodiversity conservation (criteria important to the international community), rubber agroforests contribute to achieving these two objectives more than rubber or oil-palm monocultures and much more than

rice/fallow rotations or cassava. Table 9, an abridged version of the matrix, suggests that introducing cloned rubber into agroforests significantly raises labour use and profitability and can increase returns to farmers. Wider adoption of this approach has the potential to help balance competing objectives by addressing the concerns of policy-makers to generate income and employment; by meeting the interests of smallholders to earn profits; and by improving the environment (Tomich et al., 2001).

Development agencies are increasingly targeting interventions towards poor and female farmers and want to know whether they are reaching these groups. In a review of 23 studies of factors affecting the adoption of agroforestry, Pattanayak et al. (2003) found that eight included gender as a variable. In five of these studies, male-headed households were found to be more likely to adopt agroforestry than female-headed households. However, these findings may reflect the access men have to resources and information rather than women's preferences. In central Kenya, women accounted for 60 percent of a sample of 2 600 farmers planting fodder trees (Franzel, Wambugu and Tuwei, 2003). A study in western Kenya showed that women used improved fallows and biomass transfer more frequently than men, who more often used mineral fertilizers (Figure 10) (Place et al., 2004).

Pattanayak et al. (2003) found 12 studies that assessed the effect of wealth or income on adoption of agroforestry. The relationship was positive in six and insignificant in the other six. Data from western Kenya showed that poor and non-poor households were equally likely to use improved fallows and biomass transfer to increase soil fertility (Figure 11) (Place et al., 2004).

LESSONS LEARNED, CHALLENGES AND OPPORTUNITIES

Much has been learned about how to promote agroforestry and increase benefits to farmers and others through research, extension and policy reform. Whereas this chapter has focused on success stories, failures have also provided important lessons. For example, the effectiveness of alley farming practices to improve soil fertility

TABLE 9

Abridged matrix: how selected land-use practices perform across criteria important to different stakeholders in Sumatra, Indonesia

STAKEHOLDERS	International community		Agriculturists	National policy-makers		Smallholders
CRITERIA	Global environmental quality		Plot level production sustainability	Social profitability	Employment	Production incentives
MEASURED BY	Carbon sequestration: time averaged (Mg/ha)	Biodiversity: plant species per standard plot	Rating	Returns to land at social prices (Rp 1 000/ha)	Labour input (days/ha/year)	Returns to labour at private prices (Rp/day)
LAND USE						
Natural forest	254	120	1	0	0	0
Rubber agroforest	116	90	0.5	73	111	4 000
Rubber agroforest with clonal planting material	103	60	0.5	234–3 622	150	3 900–6 900
Upland rice/bush fallow	74	45	0.5	53–180	15–25	2 700–3 300
Continuous cassava degrading to *Imperata* spp.	39	15	0	315–603	98–104	3 895–4 515

Note: 1 Rupiah (Rp) = US$0.00012 (2000).
Source: Adapted from Tomich *et al.*, 2001.

FIGURE 10

Use of soil fertility management options by gender of household head, western Kenya

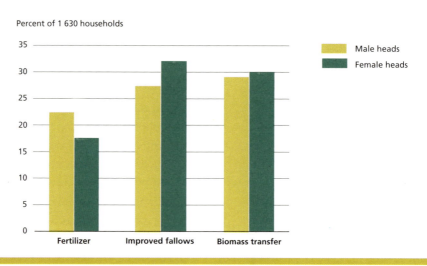

Source: Place *et al.*, 2004.

and crop yields helped refocus strategies on growing trees and crops in rotation rather than together. Some trees, such as *Leucaena leucocephala*, have become invasive in some areas, and this has helped researchers to recognize the importance of screening species.

Benefits of agroforestry

In a review of 56 agroforestry practices in 21 projects in Central America and the Caribbean, Current and Scherr (1995) found that 75 percent had positive net present values. In two-thirds of the cases, net present values and returns to labour

FIGURE 11
Use of soil fertility management options by household wealth class, western Kenya

Percent of 1 630 households

Very poor
Poor
Non-poor

Fertilizer | Improved fallows | Biomass transfer

Source: Place *et al.*, 2004.

were superior to those in alternative enterprises. In both developed and developing countries, however, agroforestry is not generally recognized as a science or a distinct practice and is rarely featured in development strategies (Garrett and Buck, 1997; Williams *et al.*, 1997). Policymakers need to be informed about the benefits of agroforestry so that they can use it to support rural development and provide environmental services (Current and Scherr, 1995). In developing countries, local authorities and traditional leaders are in a good position to promote agroforestry.

Substitutes for purchased products. Many farmers appreciate agroforestry because it generates cash income through the sale of tree products. It also provides products that the farmer would otherwise have to purchase – an important consideration, given the lack of working capital in many farming systems. For example, farmers substitute nitrogen-fixing plants for mineral fertilizers, fodder shrubs for expensive dairy meal and home-grown timber and fuelwood for wood bought off the farm.

Enhanced diversity and reduced risk. Agroforestry enhances diversity both in terms of plant biodiversity and enterprise diversity.

The latter decreases risk and allows farmers to reduce seasonal labour peaks, earn income throughout the year and accrue benefits at different times – over the short, medium and long term. Also, farmers often value trees because little effort is required to maintain them and they can be sold whenever cash is needed.

Complement to natural forest management. Evidence suggests that where farmers have incentives to plant trees and have access to information and planting material, they depend less on neighbouring forests and are less likely to damage them. Sound policies and extension programmes, as well as effective forest management mechanisms, can significantly enhance the impact of agroforestry on forest protection.

Factors affecting performance
Adaptation to local conditions. Successful efforts to introduce agroforestry often combine modern science and traditional knowledge. Experience has also shown that individual preferences, adaptations and entrepreneurial skills make a big difference and that communities need help to document and spread innovations of farmers. To minimize risk,

Faidherbia albida agroforestry/agrosilvipastoral system

One of the most pronounced agroforestry and agrosilvipastoral systems in the gum belt of sub-Saharan Africa is the one that uses *Faidherbia albida*, a tree that attains enormous size in such areas as the foothills of Jebel Marra in Darfur, Sudan. Having learned the tree phenology over centuries, communities in Darfur fence and crop the entire areas under *F. albida* with staple (sorghum and millet) and cash crops (tomato and chilli, for example).

The tree sheds its leaves during the rainy season (July to October), allowing light over the entire crown to the bole. During winter and summer (November to June), the tree produces leaves and pods that cast a heavy shade. Livestock, particularly sheep and goats, visit the tree for crop residues, shade and pods. In so doing, they add animal manure to a soil already improved through nitrogen fixed by the *F. albida* root system and the decomposition of twigs and leaflets.

F. albida usually grows along seasonal watercourses with shallow water table and is irrigated from hand-dug wells. When felled in thinning operations or when the tree is wind thrown following root collar zone rot, the wood is used in carpentry and for utensils such as mortars, oil mills and shoe lasts.

Scientists and academics need to acknowledge that today's practices and terminology have their origins in traditional knowledge and that other sound and sustainable aspects of such knowledge should be recognized and taught at all levels. Investigating myths that surround *F. albida*, including those related to the shedding of its leaves during the rainy season, might also improve understanding of current systems.

farmers prefer to choose from different options to solve a problem rather than have to rely on a single approach (Franzel and Scherr, 2002).

Availability of information and training. Farmers need more information and training for agroforestry relative to other agricultural activities, which limits the spread of some practices. When starting operations, they often lack skills to establish tree and shrub nurseries, pre-treat the seeds and carry out tree pruning activities. However, extension strategies, including field schools, exchange visits and farmer training, are effective ways of disseminating needed information.

Government and project support. Lack of financial credit is not a major constraint to adopting agroforestry practices because of the small size of farms and scale of operations, the incremental approach that farmers use to plant trees and the desire of most farmers to avoid risks. In many

instances, offering free inputs or paying farmers to plant trees encourages dependency and acts as a disincentive to planting when a project ends. Once farmers start planting on a small scale and see the benefits, they are usually able and willing to continue. On the other hand, government and project interventions are needed to promote tree planting, provide information and technical assistance and fill other gaps such as supplying tree seeds where they are not available. In most cases, however, credit or payments to farmers for planting trees are not required and may do more harm than good (Current and Scherr, 1995; Scherr and Franzel, 2002).

Linking farmers to markets. Assessing demand before planting trees is a critical first step in adopting agroforestry, as looking for a market only in times of surplus is problematic. It is also more advantageous to assist farmers to sell their produce locally before they attempt to enter a more competitive export market, and

to help them strengthen their links with the private sector as part of market development. In addition, training in entrepreneurship and business skills has proven highly beneficial to farmers, and farmer organizations can have an important role in assembling produce, bargaining collectively and reducing transaction costs.

Secure land tenure and exemptions from government ordinances. Farmers with insecure land rights are unable or unwilling to plant trees. However, formal land registration is not always necessary, as some traditional forms of tenure provide the security to plant trees (Place, 1995). A critical constraint, especially in semi-arid and arid zones, is that livestock often graze freely, feeding on or trampling newly planted trees. In some communities, restrictions now prevent this practice, and lessons need to be shared to address the problem elsewhere. In many countries, bans on cutting down trees are a disincentive for farmers to plant them. Therefore, mechanisms are needed to exempt trees on farms from such ordinances (Current and Scherr, 1995).

Decentralized, community-based germplasm strategies. The most successful approaches to supplying and distributing planting material are those involving community-based seed stands and nurseries managed by individual farmers or groups. Seed and nursery enterprises can also help to increase incomes. Efforts are needed to ensure the quality and diversity of planting material (Current and Scherr, 1995; Franzel, Cooper and Denning, 2001).

CONCLUSIONS

The proportion of trees on farms and in forests varies considerably among countries, but two trends seem almost universal in the tropics: the number of trees in forests is declining, and the number on farms is increasing. In a survey of 64 communities in Uganda, for example, the proportion of land under forest declined from 4 to 2 percent between 1960 and 1995, while that under agriculture increased from 57 to 70 percent. Interestingly, the proportion

First World Agroforestry Congress

Participants from 82 countries attended the first World Agroforestry Congress in Florida, United States, from 27 June to 2 July 2004. During discussions, they noted significant progress over the past 25 years in building a scientific foundation for agroforestry systems. Recognizing the links to the United Nations Millennium Development Goals, the congress called on countries, international organizations, the private sector and other partners to use the full potential of agroforestry:

- to increase household income;
- to promote gender equity;
- to empower women;
- to improve the health and welfare of people;
- to promote environmental sustainability.

Experts further noted the need to increase investments for research, technology development and extension so as to integrate agroforestry more fully with natural resource and watershed management. They also urged governments to highlight the role of agroforestry in poverty reduction strategies, to provide financial support and to develop policies that promote the adoption of associated practices.

of agricultural land under tree cover increased from 23 to 28 percent (Place, Ssenteza and Otsuka, 2001).

Agroforestry has made tremendous strides in recent years, but many challenges remain in terms of its wider application. It is necessary to identify and measure the range of benefits, given that they are not well documented. Moreover, additional research is required to quantify the benefits to various stakeholders, to deal with the variability in benefits, to assess the effects and trade-offs of different policies and to examine the impact of agroforestry practices on forest protection, particularly in the tropics. Determining which practices are most suited to particular groups, such as women

Economics of wood energy

In the past decade, policies to encourage the use of renewable energy have grown in importance as part of the efforts to reduce dependence on non-renewable energy sources such as fossil fuels and as part of strategies to address global warming. Wood energy has been identified as a potentially significant source of renewable energy, and for this reason a number of developed countries have shown interest in increasing its use (Trossero, 2002). In addition, wood energy remains the most important source of energy for the more than two billion people in developing countries who have access to few other sources of energy.

Given the importance of wood energy in developing countries and its potential importance in developed countries, it is useful to understand the economic forces that encourage or constrain the use of wood energy. This chapter provides an overview of wood energy and its importance, explains some of the economic forces affecting wood energy production and consumption and describes how countries might develop the wood energy sector to meet some of their broader policy goals and objectives.

OVERVIEW OF WOOD ENERGY

Wood energy comprises a number of different types of wood-based fuels. The most prominent of these is fuelwood, cut directly from trees and forests. This may be further refined into other types of energy such as charcoal or wood-derived liquid fuels. In addition to these, wood energy includes a number of by-products from the forest processing industry (notably black liquor – a by-product of pulp and paper making – and wood residues) and recycled wood and paper. It should also be noted that the wood energy sector includes more than just fuelwood and charcoal.

Currently, wood energy accounts for about 5 percent of the world's total primary energy supply (TPES)[1], and woodfuel is by far the most important source of wood energy (Figure 12). However, the importance of wood energy to total energy supply differs greatly among countries and regions. For example, wood energy (mostly fuelwood) accounts for more than two-thirds of TPES in the Congo, Eritrea, Ethiopia, Mozambique and the United Republic of Tanzania, and it accounts for over half of TPES in Haiti, Nepal and Paraguay. In Europe, the overall contribution of wood energy to TPES is very low (around 1 percent), but there are great differences among countries. For example, because of the large pulp and paper industry and the use of black liquor for energy production, wood energy accounts for 14 and 10 percent of TPES in Finland and Sweden, respectively (Table 10).

The importance of wood energy as a use of forests and trees also varies widely among countries and regions. Overall, woodfuel (i.e. fuelwood and charcoal) accounts for about 53 percent of total roundwood produced in the world. However, woodfuel accounts for only 14 percent of total production in G8 countries, compared with 69 percent in the rest of the world (Table 11). In terms of the distribution of woodfuel production across regions, Asia accounts for the largest share of global woodfuel production (around 44 percent), followed

---

[1] Total primary energy supply is the supply of unprocessed fuels (e.g. oil, gas and coal) and excludes the production of refined or converted types of energy (e.g. petrol and electricity). The figures presented here have been calculated by converting all of the different types of fuel into comparable measures of the energy that they can produce.

FIGURE 12
Contribution of wood energy to total primary energy supply, 2001

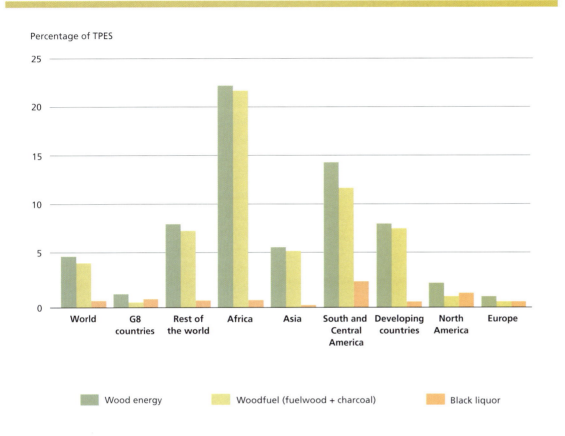

Percentage of TPES

Wood energy   Woodfuel (fuelwood + charcoal)   Black liquor

Source: International Energy Agency, 2003.

TABLE 10
Contribution of wood energy to total primary energy supply
in selected developed countries, 2001

Country	Contribution to TPES (%)	
	Black liquor	All wood energy
Finland	11.5	14.4
Sweden	8.0	9.9
Canada	3.0	3.5
New Zealand	2.0	2.0
United States	1.3	2.0

Source: International Energy Agency, 2003.

TABLE 11

Percentage of total roundwood production used for woodfuel, 1997

Region	Proportion of total roundwood production (%)
World	53
G8	14
Rest of the world	69
Developing countries	76
Africa	89
Asia	79
Europe	18
North America	15
South and Central America	59

Source: FAO, 2004.

by Africa (21 percent). Together, Asia, Africa and South and Central America account for 76 percent of global woodfuel production (Trossero, 2002).

In the future, the global production of woodfuel is expected to increase moderately, from 1 885 million cubic metres in 2000 to 1 921 million in 2010 and 1 954 million in 2020 (Broadhead, Bahdon and Whiteman, 2001). Fuelwood production is expected to increase in Africa and South America but decrease in Asia, while all three regions are expected to increase charcoal production. In addition, the use of black liquor for energy is likely to increase in countries where the pulp and paper industry is expanding.

ECONOMIC VALUE OF WOOD ENERGY PRODUCTION

Wood energy contributes directly to national economies as a source of energy supply. However, because a large proportion of wood energy is not sold in the market, valuing this contribution is quite difficult. In addition, the social and environmental impacts of wood energy production and consumption are indirect effects – or externalities – of wood energy use. These externalities can be positive

or negative and are also very difficult to value.

The contribution of any activity to the economy (e.g. to the GDP) is measured as the value added produced by that sector. This in turn is calculated by subtracting the costs of goods and services purchased from other sectors and used for production (e.g. fuel, tools and machinery) from the total value of output in the sector (i.e. quantity produced multiplied by price). The production of woodfuel involves few purchases from other sectors. This is particularly true in developing countries, where the main input used to produce woodfuel is labour (which is not counted as a cost in the calculation of value added). Thus, the total value of woodfuel production gives a reasonable approximation of the value added in the sector.

Currently, woodfuel prices range from around US$5 to $25 per cubic metre in developed countries and US$1 to $10 per cubic metre in developing countries (Broadhead, Bahdon and Whiteman, 2001). However, in developing countries, a large proportion of woodfuel is produced by individuals for their own consumption rather than for sale. In such cases, there are several ways of valuing production that is not traded in the market. One is to calculate the replacement cost of this production (i.e. the cost of replacing the production for personal use by the purchase of woodfuel or other types of energy), but this is likely to lead to an overestimate of the value of production. Alternatively, the value of production can be calculated as the cost of the time taken to collect woodfuel (as the value must be at least equal to this cost or collection would not take place), but this would probably lead to an underestimate of the value of production.

Taking into account these uncertainties, the market price of woodfuel can be used as a rough estimate of the value of woodfuel production. Therefore, with total production of around 1 885 million cubic metres (and assuming 75 percent in developing countries and 25 percent in developed countries), the total value of global woodfuel production could be in the range of

US$4 billion to $26 billion per year. These figures amount to about 0.01 to 0.06 percent of global GDP. Other types of wood energy (e.g. black liquor) are not included here, so these figures are an underestimate. However, they indicate that the direct contribution of wood energy to national economies is probably quite small.

Positive and negative externalities

The main positive externalities of wood energy are the effect on carbon balances of substituting wood energy for fossil fuels and the employment generated by wood energy production. The main negative externality is the environmental cost of woodfuel harvesting in terms of forest loss and degradation.

With the methodology currently used for carbon accounting, losses of biomass carbon are counted as part of changes in the stock of forest biomass. Thus, to avoid double counting, the use of wood energy is not counted as an activity that leads to CO_2 emissions even though it is one.

The potential for wood energy to lead to real changes in carbon balances depends on the source of woodfuel. If woodfuel is produced from sustainably managed forests where the wood harvested is replaced by the increment of the remaining growing stock, then the substitution of wood energy for fossil fuels will result in a real reduction in the net carbon balance. Similarly, if residues from harvesting and wood industry are used for energy production rather than left for waste, this would also have a positive net effect.

However, if woodfuel is produced on an unsustainable basis by clearing forests, the substitution of wood energy for fossil fuels will not have a positive effect on carbon balances and could even be worse than the use of fossil fuels. This is particularly true if wood energy is produced inefficiently. For example, inefficient kilns emit a great deal of CO_2 during the production of charcoal, resulting in very high emissions per unit of energy produced.

With respect to employment, woodfuel production is labour intensive and an important source of income and employment for rural households. Woodfuel production requires the highest amount of labour inputs per unit of energy produced: 100 to 170 person-days per terrajoule for fuelwood and 200 to 350 person-days per terrajoule for charcoal (Remedio, 2001). However, the benefit of this employment generation depends on the value of the labour used for production (Luoga, Witkowski and Balkwill, 2000). For example, employment can be considered as a positive externality if rural unemployment is high, but perhaps not if there are alternative uses for this labour. In addition, policy-makers should be aware that woodfuel projects and programmes may not always be the best way to increase rural income and employment.

As with the impact on carbon balances, the environmental costs of wood energy use also depend on the source of woodfuel. Again, forests that are sustainably managed for woodfuel production are likely to lead to some positive externalities in terms of their environmental impact, while unsustainable harvesting for woodfuel production is likely to lead to environmental costs.

To summarize, the indirect effects of wood energy production and consumption are complex and not well known. However, it seems likely that on balance there may be some positive externalities from the use of wood energy in developed countries and negative externalities in many developing countries.

ECONOMICS OF WOOD ENERGY PRODUCTION AND CONSUMPTION IN DEVELOPING COUNTRIES

In developing countries, the use of wood energy is divided as follows: fuelwood, 90 percent; black liquor, 6 percent; and charcoal, 4 percent. Households (particularly rural households) are its main consumers, with industry and the service sector consuming far less.

The use of wood energy is determined by a number of factors, including price, income, availability of other types of energy and resource availability. In general, most consumers in developing countries use wood

Many rural households in developing countries only produce enough woodfuel to meet their own needs; some, however, are able to enter the market as sellers of woodfuel

FAO/17437/A. ODOUL

energy because their choice of energy supply is restricted by income and the availability of other types of energy.

Households that use wood energy can be divided into four types:

- households that only produce woodfuel for their own needs;
- households that produce and sell woodfuel;
- households that produce and purchase woodfuel;
- households that only purchase woodfuel.

Most rural households fall into the first and second groups, while most urban households fall into the third and fourth groups.

The price of woodfuel has a greater effect on consumption for the last three groups in the above list. For example, households that only purchase woodfuel are likely to respond to price changes by altering total energy consumption or switching to other types of energy. Price changes are likely to influence total production for the second group or total consumption for the third group. The effect of price changes on how much these groups produce for themselves will depend on the value of the labour they

expend to produce woodfuel. For example, if prices rise, households in the third group are likely to produce more of their own woodfuel. In most cases, households in the first group do not participate in the market for reasons such as location (i.e. remoteness) and the low value of their own labour. However, if woodfuel prices change significantly, households in this group could enter the market either as buyers or sellers of woodfuel.

With respect to income, some researchers have found that the share of woodfuel in household energy use declines as per capita income increases (Sathaye and Tyler, 1991; Leach, 1988; Broadhead, Bahdon and Whiteman, 2001). On the other hand, Leach *et al.* (1986) reported that woodfuel consumption increased when incomes increased in very poor rural households in Brazil, India, Pakistan and Sri Lanka. Some others have also reported a positive relationship between income and woodfuel consumption (Shaw, 1995; Zein-Elabdin, 1997). Thus, it is not always the case that low-income households first use woodfuel and then ultimately progress to other types of energy as incomes increase. High-income households may consider woodfuel an inferior good, but low-income households may not share this view. Consequently, in poor countries, the switch from woodfuel to other types of energy is likely to occur slowly.

Generally, the decision to switch depends on the prices, availability, reliability of supply and energy content of the alternatives. Another factor is the cost of changing equipment (e.g. stoves). However, in many rural areas, there is no option but to use woodfuel because of remoteness and the lack of infrastructure for delivering other types of energy.

Surprisingly, black liquor contributes a little more than charcoal to the TPES of developing countries, but this is the result of high use of black liquor in only a few countries where pulp and paper production is significant (e.g. Brazil, Chile, China, Colombia, Indonesia and South Africa). The availability of by-products from the forest processing industry and recycled wood and paper products is significant and could be used to increase wood energy production, but this will depend on the profitability of using these materials for energy production compared with the profitability of alternative uses (e.g. as inputs for wood panel and paper manufacturing).

Other social and environmental factors that affect household woodfuel consumption include climate (e.g. altitude, length of winter and rainy seasons), access to markets and forest resources, health and environmental effects of woodfuel use (e.g. smoke) and cultural variables. For example, the failure of fuelwood and charcoal substitution programmes in many countries is attributed to the resistance of consumers to change cooking habits (e.g. to replace wood and charcoal stoves with alternative technologies). These other factors can be important and should be considered in wood energy policies and programmes.

ECONOMICS OF WOOD ENERGY PRODUCTION AND CONSUMPTION IN DEVELOPED COUNTRIES

With few exceptions, black liquor is the main type of wood energy used in developed countries. In 2001, black liquor accounted for

0.9 percent of TPES in G8 countries, compared with a total of 1.4 percent for all wood energy (Figure 12 and Table 10). In the countries of OECD, the contribution of all biomass energy to TPES is about 3.5 percent; biomass energy from agriculture and forestry accounts for about 86 percent of this (Radetzki, 1997).

Government efforts to boost the production of renewable energy include attempts by the EU to increase the share of renewable energy to 12 percent of all energy consumption and 22 percent of electricity consumption by 2010. In the EU plan, the expected growth in biomass energy production is second highest (after wind power), with an expected increase from 55 million to 135 million tonnes of oil equivalent (Harmelink *et al.*, 2004). Most developed countries treat biomass as an important source of renewable energy and have supporting policies in place (Table 12). In addition to governments, many other organizations also promote renewable energy. However, despite such initiatives, concerns remain about the production costs and financial viability of renewable energy production.

The cost of wood energy production depends on the source of wood used. In general, recovered wood and paper products and wood residues from the forest processing industry are likely to be the least costly sources of supply because they are concentrated in urban areas and can benefit from economies of scale in production. Harvesting residues and the forest plantations specifically managed for wood energy production are likely to be more expensive sources of supply. Consequently, wood energy systems in developed countries have tended to focus on using wood residues. However, there is an opportunity cost of using these materials for wood energy, as they are also an important source of supply for the forest industry. Thus, there are concerns about the impact that subsidizing wood energy will have on the forest industry. Promoting wood energy will be beneficial for the forestry sector as a whole, but the distribution of the costs and benefits of such policies across the sector needs to be evaluated carefully.

TABLE 12
Instruments used in OECD countries to promote renewable energy

	Austria	Belgium	Denmark	Finland	France	Germany	Greece	Ireland	Italy	Japan	Luxembourg	Netherlands	Norway	Portugal	Spain	Sweden	UK	USA
Research and development	■	■	■				■		■		■	■				■		■
Tax incentives		■	■	■	■				■		■	■		■		■		■
Subsidized loans	■	■			■	■	■	■		■	■	■		■	■			
Capital subsidies	■	■	■			■	■	■	■			■		■	■	■		
Feed-in tariffs	■	■	■		■	■	■		■		■	■		■	■	■		
Energy taxes	■											■						
Market liberalization						■		■					■			■		■
Information campaigns	■	■		■				■		■	■	■					■	■
Training				■				■	■			■	■					
Standardization			■		■			■		■	■	■	■					
Certification					■			■				■						

Source: Short and Keegan, 2002.

Green pricing programmes for renewable energy

In 2002, 90 green pricing programmes were offered to about 26 million consumers in 32 states in the United States. Almost 274 000 consumers participated in these programmes. The premiums for renewable energy ranged from US$0.007 to US$0.176 per kilowatt-hour, and consumers paid an average of US$4.43 per month for green power.

At the end of 2002, utility companies had installed nearly 290 megawatts of renewable energy capacity and had plans to install another 140 megawatts. Biomass energy production accounted for the second largest share of capacity, with 15 percent of installed capacity and 17 percent of planned capacity. About 25 percent of utility companies produced their own renewable energy, 46 percent purchased all of their supplies from other power generators or purchased renewable energy certificates, and the remaining companies used a combination of these approaches.

Source: Bird, Swezey and Aabakken, 2004.

More fuel-efficient stoves improve the welfare and living conditions of people living in remote communities

Other factors that will affect the economic viability of wood energy are the demand for renewable energy and the non-wood costs of wood energy production. With respect to demand, energy pricing programmes in some developed countries have enabled consumers to choose renewable energy and pay slightly more for it (see Box on facing page). In addition to households, corporate consumers in the industrial and service sectors are starting to purchase renewable energy to improve their environmental image and as part of social corporate responsibility programmes. Thus, the prices for renewable energy may increase in the future, particularly if the market can be divided in this way.

In terms of production costs, the current cost of electricity production from biomass is about US$0.07 to $0.09 per kilowatt-hour, which is slightly higher than the cost of producing electricity from fossil fuels. However, in favourable situations, it can be reduced to as little as US$0.02 to $0.04 per kilowatt-hour (Ahmed, 1994). Furthermore, new and improved technologies, such as integrated biomass gasification plants, may soon produce electricity from biomass at about US$0.04 per kilowatt-hour (Elliott, 1993). More generally, Short and Keegan (2002) predict that the cost of biomass energy production will fall by 15 to 20 percent over the next 20 years, making it broadly comparable with the cost of energy from fossil fuels.

FUTURE STRATEGIES AND POLICIES

Over the next two decades, the importance of wood energy in developed countries is likely to increase as part of efforts to promote the use of renewable energy. This may also occur in developing countries, although the greatest changes can be expected from households switching from woodfuel to other types of energy. These transitions will require programmes and policies that take into account the complex economic forces that influence wood energy production and consumption. The following issues are put forward for consideration by policy-makers.

- At the international and national level, forestry and energy policies need to be complementary in order to achieve the benefits that wood energy can offer.
- Government subsidies for wood energy should continue in order to enable it to compete with other types of energy. However, subsidies need to take into account the impacts of greater wood energy use on other parts of the forestry sector.
- Policies and projects that encourage the use of wood energy should be based on holistic analysis of all the economic, social and environmental costs and benefits of wood energy. In situations where the use of wood energy results in significant benefits, this information should be disseminated widely.

- Attention should be paid to possible negative externalities of woodfuel use (such as nitrogen oxides and particulate emissions), which are largely unknown at the moment.
- Efforts should continue to improve the efficiency of wood energy production in developing countries. These could include not only the promotion of more efficient wood stoves but also the development of more modern production systems such as the use of wood for electricity production. Successful experiences with modern wood energy systems in some developed countries should be shared with developing countries through investment and technology transfer.
- Integrated operations that combine the use of wood for energy and the production of forest goods are likely to be more economically viable than systems that only focus on the production of wood energy. ◆

REFERENCES

Ahmed, K. 1994. *Renewable energy technologies: a review of the status and costs of selected technologies.* Washington, DC, World Bank.

Bird, L., Swezey, B. & Aabakken, J. 2004. *Utility green pricing programs: design, implementation and consumer response.* Golden, USA, National Renewable Energy Laboratory.

Broadhead, J., Bahdon, J. & Whiteman, A. 2001. *Past trends and future prospects for the utilization of wood for energy: Annexes 1 and 2.* Global Forest Products Outlook Study Working Paper No. GFPOS/WP/05. Rome, FAO.

Elliott, P. 1993. Biomass energy overview in the context of the Brazilian biomass power demonstration. *Bioresource Technology,* 46: 13–22.

FAO. 2004. *Wood energy data from the Energy Information Systems.* Rome (available at www.fao.org/forestry/site/14012/en).

Harmelink, M., Voogt, M., Joosen, S., Jager, D., Palmers, G., Shaw, S. & Cremer, C. 2004. *Implementation of renewable energy in the European Union until 2010.* Utrecht, Netherlands, Ecofys.

International Energy Agency. 2003. *Key world energy statistics 2003.* Paris.

Leach, G. 1988. Residential energy in the third world. *Annual Review of Energy,* 13: 47–65.

Leach, G., Jarass, L., Obermair, G. & Hoffman, L. 1986. *Energy and growth: comparison of 13 industrial and developing countries.* Guildford, UK, Butterworth Scientific.

Luoga, E.J., Witkowski, E.T.F. & Balkwill, K. 2000. Economics of charcoal production in miombo woodlands of eastern Tanzania: some hidden costs associated with commercialization of the resources. *Ecological Economics,* 35: 243–257.

Radetzki, M. 1997. The economics of biomass in industrialized countries: an overview. *Energy Policy,* 25(6): 545–554.

Remedio, E.M. 2001. *Socio-economic aspects of bio-energy: a focus on employment.* Rome, FAO. (Unpublished)

Sathaye, J. & Tyler, S. 1991. Transition in household energy use in urban China, India, the Philippines,

Thailand, and Hong Kong. *Annual Review of Energy and Environment*, 16: 295–335.

Shaw, C.L. 1995. New light and heat on forests as energy reserves. *Energy Policy*, 23(7): 607–617.

Short, W. & Keegan, P. 2002. The potential of renewable energy to reduce carbon emissions. *In* R.G. Watts, ed. *Innovative energy strategies for CO$_2$ stabilization*, pp. 123–177. Cambridge, UK, Cambridge University Press.

Trossero, M.A. 2002. Wood energy: the way ahead. *Unasylva*, 211: 3–12 (also available at www.fao.org/forestry/unasylva).

Zein-Elabdin, E.O. 1997. Improved stoves in sub-Saharan Africa: the case of Sudan. *Energy Economics*, 19: 465–475. ◆

Tariffs and non-tariff measures in trade of forest products

The forest products sector is estimated to contribute about 1.2 percent of world GDP and approximately 3 percent of international merchandise trade. Industry annual turnover exceeds US$200 billion for four product categories: roundwood and sawnwood, panels, pulp, and paper. In 2003, global production of industrial roundwood was close to an estimated 1.6 billion cubic metres, with an increasing proportion of timber originating from planted forests. Forests also provide important goods and services, including wood energy, food and other non-wood products, for 1.2 billion people of whom approximately 90 percent live below the poverty line (FAO, 2004a).

Trade in industrial roundwood has doubled in volume over the past 40 years and is expanding rapidly. Trade in processed products is increasing as well. While exports of forest products from non-tropical countries are on the rise, those from tropical countries seem to have remained fairly static since the 1970s (FAO, 2004b). In terms of markets for forest products, Europe is experiencing growth, markets in the United States continue to be strong, and China has become one of the world's largest log and sawnwood importers, particularly of tropical timber (UNECE/FAO, 2003).

Although global trade in forest products is expanding, it is influenced by trade measures that determine market access and vary considerably by product, region and country, including:

- import tariffs;
- export restrictions, including logging bans;
- technical product standards, including production and processing methods;
- sanitary and phytosanitary measures;
- environmental and social standards, including certification and product labelling.

Import and export tariffs and most non-tariff measures are based on national policies and legislation. Concerns over forest degradation and loss of forest cover, however, are heightening pressure on governments, the private sector and international institutions to address the impact and interaction between trade and the environment, and specifically their relation to sustainable forest management (ITTO, 2003). Deliberations on international and regional trade, including those of the Committee on Trade and Environment (CTE) of the World Trade Organization (WTO), are focusing on these issues. Thus, there are indications that the obligations countries assume when they become members of WTO and regional trade agreements will increasingly affect the terms of trade in forest products and services (Neufeld, Mersmann and Nordanstad, 2003).

IMPORT TARIFFS AND TARIFF ESCALATION: ATTEMPTING TO MEET INTERNATIONAL OBLIGATIONS

As a means to target market access and market shares for domestic producers of timber and wood-based products, tariff escalations – higher tariffs applied to the import of value-added products – are widely used to support and protect domestic forest industries and small-scale producers. Where the forest sector is evolving, many governments also offer subsidies and other incentives for forest production and processing (Rytkönen, 2003).

While the Uruguay Round of trade negotiations brought about significant reductions in import tariffs, tariff escalations still limit trade in forest and wood-based products. Moreover, the fact that some large importers such as China did not participate in the Uruguay Round diminishes the impact of its outcomes. In developed countries, tariffs on forest and wood-based products are low – less than 5 percent for most – and have a limited effect on imports. Significant exceptions for some countries and products apply to wood-based panels and paper products that have tariff rates between 10 and 15 percent (UNCTAD, 2003a).

Tariff rates are more elevated in developing countries, particularly in Asia where they usually range between 10 and 20 percent but can be considerably higher (FAO, 2004a). On the other hand, generalized systems of preferences (GSPs) and special arrangements under regional and bilateral trade agreements reduce the effects of tariffs on imports into developed countries. These tariffs, particularly on value-added products, were established in many countries to support domestic industrialization rather than to support sustainable forest management. However, recent efforts to link tariff measures and environmental issues are meeting with some criticism. An example is the GSP of the EU, which grants a tariff preference for selected tropical timber products from sources that are managed according to internationally acknowledged standards and guidelines (Council of the European Union, 2003).

NON-TARIFF MEASURES: CAPTURING THE POTENTIAL

Non-tariff measures, unlike tariffs, are difficult to characterize. It is not easy to determine whether a non-tariff measure is government initiated or consumer driven. It is also difficult to ascertain whether a non-tariff measure has been put in place to support and protect domestic forest production and industries or whether it was introduced to support sustainability within the forest sector and its industries. Concerns of consumers and civil society groups, mostly in developed countries, often coincide with government objectives in applying non-tariff measures (Borregaard and Dufey, 2001). In many cases, such action is taken to enhance sustainable forest management, particularly in the tropics. However, non-tariff measures also cover policy decisions that, on the surface, are not connected to trade and market development for forest products – for example, support to markets for environmental services (Shahin, 2002).

In contrast to regulatory mechanisms, many non-tariff barriers to trade are informal, consumer based and government supported. The best examples are forest certification (see Box on page 110) and product labelling schemes, which exporting countries often perceive as trade barriers because of their potentially significant impact on both volume of trade and product composition. Although market based, these schemes influence the development of national policies and government actions such as standard setting for forest management and wood processing. As a result they are subject to controversial debate at all levels.

Empirical evidence shows that non-tariff measures that are driven by environmental and social concerns frequently limit market access, especially to tropical timber (Rytkönen, 2003). Legislation developed by the EU and other consumer countries to restrict imports of illegally harvested and traded timber is an example of such a measure (Council of the European Union, 2003). As a result, trade patterns and geographic distribution of trade in forest and wood-based products tend to shift towards less sensitive markets where concerns over forest management and production and processing methods have less influence on market access and market shares (Sun, Katsigris and White, 2004).

In producer countries, national export restrictions continue to be among the more frequently applied non-tariff measures of significance. They include total export bans, export quotas and selective bans based on species; direct charges such as export taxes or export levies; restrictions on quantity because

Certified forest worldwide

The area of certified forest has been increasing steadily (Figure 13). Close to 90 percent of the more than 176 million hectares certified around the globe are situated in the Commonwealth of Independent States, Europe and North America. However, this amount represents less than 4 percent of the world's forests.

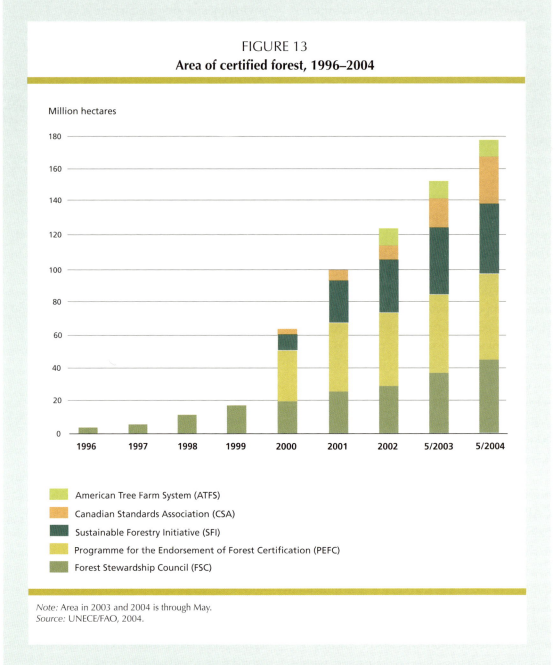

FIGURE 13
Area of certified forest, 1996–2004

Million hectares

Legend:
- American Tree Farm System (ATFS)
- Canadian Standards Association (CSA)
- Sustainable Forestry Initiative (SFI)
- Programme for the Endorsement of Forest Certification (PEFC)
- Forest Stewardship Council (FSC)

Note: Area in 2003 and 2004 is through May.
Source: UNECE/FAO, 2004.

of limits on harvest levels; and administrative controls such as permits and licences. Export restrictions are common in most developing, and in some developed, countries. Compliance with CITES constitutes a trade limitation for those forest products listed in the agreement's annexes (Mulliken, 2003).

In the past, countries exporting tropical timber used export taxes to raise revenue and support domestic industries. Taxes ranged from 10 to 20 percent for logs, while those for processed products such as veneer and plywood were often negligible because of the need to promote forest-based commercial activities (FAO, 2004a). In cases where export taxes are low and represent the only means for governments to capture income, they should not be regarded as a way to discourage exports. Policy objectives, however, have generally shifted towards investment incentives so that export restrictions, including total bans of logs and sawnwood, have replaced export taxes. Although sometimes criticized, such restrictions can contribute to industrial development and prevent the destruction of forests, albeit at a substantial cost. They can also enhance people's well-being, provided that the restrictions are adapted to local situations and used in combination with other policy instruments aimed at rural or industrial development (Hoekman and Kostecki, 2001).

POLICIES THAT AFFECT TRADE AND NON-TARIFF MEASURES
Incentives
Incentives, including subsidies, are common in the forest sector to promote timber production, reforestation and investments in natural and planted forest management where returns are too low to attract private financing (Schmidt, 2003). Incentives have also been used for strategic reasons, for example to build sufficient wood supply to encourage processing ventures. From a trade policy perspective, financial subsidies to promote production particularly influence the competitiveness of individual producers, although such incentives

often result in excessive harvesting and are a concern to governments and some segments of society (ITTO, 2003).

The WTO Agreement on Subsidies and Countervailing Measures contains special provisions for developing countries, including exemptions from prohibiting export subsidies in countries with a per capita gross national product (GNP) of less than US$1 000 per annum. Among permissible incentives are those that are non-specific with regard to industrial unit or sector and those that support research and development in disadvantaged regions or for environmental purposes. Injured importing countries can impose countermeasures and countervailing duties in certain circumstances (WTO, 2003).

Some constituencies believe that low concession fees and undervalued timber result in underpricing of forest resources, especially in tropical areas, and see this underpricing as a subsidy and as one of the main causes of forest degradation. While WTO may not find that such incentives contravene its provisions, these and similar types of assistance are of international concern. Agricultural subsidies that make crop production or grazing an attractive land use have often caused adverse effects on forests. Intended to foster rural development and sustainable livelihoods, they have proved to be unsustainable if applied without considering environmental, social and ecological implications, including those related to climate, water, erosion control and biodiversity (UNCTAD, 2003b).

Incentives such as grants, tax concessions and schemes to promote research and sustainable forest practices that are linked to the environmental and social services of forests are increasingly finding acceptance. To promote sound management further, especially in developing countries, targeted and temporary incentives may be required. The development of forest policies that conform to international and regional obligations will prevent incentives from becoming a trade issue (FAO, 2004a).

A shipment of logs that have not been debarked, illegally offloaded in the Seychelles for transhipment elsewhere, violates phytosanitary regulations necessary to stem the threat of pest introductions – which could become one of the biggest hidden environmental costs of trade

Technical barriers to trade

The objectives of the WTO's Agreement on Technical Barriers to Trade (TBT) are to ensure that technical regulations and standards are not used as disguised protectionist measures and that products from members of WTO are not treated less favourably than like products of national origin; and to reduce the extent to which technical regulations and standards operate as barriers to market access. The following provisions are examples of the TBT Agreement regulations regarding standard setting.

- International standards are to be used if they exist and are relevant.
- National standardizing bodies shall participate in preparing international standards.
- National standardizing bodies shall avoid duplication of, or overlap with, the work of other national, regional and international standardizing bodies.
- Every effort shall be made to achieve national consensus on standards.
- The standardizing body shall specify standards based on product requirements in terms of performance rather than design or descriptive characteristics.

The TBT Agreement sets out procedures to ensure that technical regulations and standards, including packaging, marking and labelling requirements, do not create obstacles to international trade. However, the text is not explicit with regard to voluntary certification and labelling. At issue are the motives behind these schemes – whether they are protectionist in intent or constitute arbitrary discrimination. While certification of forest products was initiated as a market-based instrument, promotion of this approach by governments and civil society is resulting in closer links' being established with national and international standard setting and policy development (WTO, 2003).

Sanitary and phytosanitary measures

The WTO Agreement on the Application of Sanitary and Phytosanitary Measures (SPS) was established to guarantee that the producer has cleaned, sanitized, sterilized or by other means rendered the offered commodity free from unwanted dirt, seeds, pests or germs. Introduced pests and diseases pose a threat to the forest resource and its biodiversity and could become one of the biggest hidden environmental costs of trade. Standards related to plant health are generally acknowledged as legitimate, as pests and diseases can have devastating effects on domestic forests. Countries may take measures that are stricter than international standards if such action is justified or if it is in response to a prescribed risk assessment. The complexity and severity of requirements and the manner in which they are enforced can have such a substantial effect on trade that some exporting countries consider them to be significant and unnecessary obstacles to trade.

Although national regulations governing sanitary and phytosanitary measures can distort trade, controls are not usually seen as unreasonable if they are scientifically based, given the need to protect human, animal or plant life or health.

TRADE AND SUSTAINABLE FOREST MANAGEMENT

The justification and effectiveness of non-tariff measures are under continuous discussion and are becoming increasingly important in negotiations related to social, environmental and economic issues. The compatibility of national policies, legislation and forest management with WTO rules is also the subject of much debate (Toyne, O'Brian and Nelson, 2002).

To date, deliberations cover more than just disputes over the application of non-tariff measures to stimulate the export of particular products or to protect domestically produced goods and national industries from foreign competition. Rules governing global trade have been established to support sustainable development as well.

WTO multilateral agreements such as the TBT and SPS Agreements are not specific to the forest sector but have important implications on how WTO member countries regulate the

trade of forest products and services. In the Doha Declaration of the Fourth Ministerial Conference in 2001 (WTO, 2001) and in the Doha Development Agenda (WTO, 2004), forest issues arise in connection with subsidies, the environment, environmental goods, ecolabelling, certification, plant health, intellectual property rights, development, market access, technical standards and regulations. Such deliberations serve to strengthen the positive interaction between trade and the move towards sustainable forest management. By the same token, discussions in CTE and elsewhere aim to clarify the relationship between WTO rules and special trade obligations in multilateral environmental agreements, including CITES, CBD, UNFCCC and ITTA (WTO, 2003).

IPF (1997) and IFF (2000) concluded that trade can have both a positive and negative impact on sustainable forest management and thus recommended that countries monitor the effects of trade policies more closely. Recently, trade in wood products derived from illegal logging has been cited as an underlying cause of forest degradation, one that negatively affects market access and market share of products from sustainably managed forests. As a result, calls are being made to take immediate action on domestic forest law enforcement and illegal international trade (see also page 48).

CONCLUSIONS

In 2003, the United Nations Conference on Trade and Development (UNCTAD) underscored that "… tariff escalation biases exports towards unprocessed resource-based commodities, characterized by low value added. This may cause difficulties to commodity-dependent developing countries in their efforts to diversify their export base". Trade in industrial roundwood is increasing rapidly worldwide, but exports, including timber from planted forests, have decreased in the tropics – a situation that reflects the impact of export bans and other non-tariff measures, increased domestic demand and declining supplies. In attempting to diversify their forest products,

developing countries and countries with economies in transition need to identify national incentives that take into account the current and future resource base, community development and the capacity of the private sector, including capital investment. They may wish to draw upon the experiences of other countries that have succeeded in developing domestic policies yet have complied with trade rules at the same time.

Certification of forest management and labelling of forest products increase access for sustainably produced wood-based products in certain markets. While certified forests, including planted forests, constitute only 4 percent of total forest area and certified forest products account for only a fraction of overall trade, producers and consumers no longer perceive certification of forest management as a negative non-tariff measure. Rather, certification schemes are being regarded as effective in improving the link between trade and forest management even though complaints continue over market access and market shares, particularly of forest products from tropical regions (Contreras-Hermosilla, 2003).

Trade measures are being changed and adjusted to respond to specific production and market situations, with most staying within the boundaries of global and regional trade agreements. Those that stem from concerns over sustainability in the forest sector will continue to be evaluated against special trade obligations in multilateral environmental agreements and against global and regional trade rules. ◆

REFERENCES

Borregaard, N. & Dufey, A. 2001. *Effects of foreign investment versus domestic investment on the mining and forestry sectors in Latin America.* Organisation for Economic Co-operation and Development. (Unpublished)

Contreras-Hermosilla, A. 2003. *Current state of discussion and implementation related to illegal logging and trade in forest products.* Rome, FAO.

Council of the European Union. 2003. *Proposal for a Regulation of the European Parliament and of the Council amending Council Regulation (EEC) No 2913/92 establishing the Community Customs Code.* Brussels.

FAO. 2004a. *Trade and sustainable forest management – impact and interactions.* Main Analytic Study of GCP/INT/775/JPN. Rome.

FAO. 2004b. *FAOSTAT database.* Rome.

Hoekman, B.M. & Kostecki M.M. 2001. *The political economy of the world trading system: the WTO and beyond.* New York, USA, Oxford University Press.

IFF. 2000. *Report of the Ad Hoc Intergovernmental Forum on Forests on its fourth session.* E/CN.17/2000/14.

IPF. 1997. *Report of the Ad Hoc Intergovernmental Panel on Forests on its fourth session.* E/CN.17/1997/12.

ITTO. 2003. *Market access of tropical timber.* Report submitted to the International Timber Council at its 33rd Session. Yokohama, Japan, International Tropical Timber Organization.

Mulliken, T. 2003. *The role of CITES in international trade in forest products - links to sustainable forest management.* Cambridge, UK, TRAFFIC International.

Neufeld, R., Mersmann, C. & Nordanstad, M. 2003. *Current state of debate in WTO on market access, technical barriers to trade (TBT) and impact of multilateral environmental agreements.* Rome, FAO.

Rytkönen, A. 2003. *Market access of forest goods and services.* Rome, FAO.

Schmidt, R. 2003. *Financial investment in sustainable forest management – status and trends.* Rome, FAO.

Shahin, M. 2002. Trade and environment: how real is the debate. *In* K.P. Gallagher & J. Werksman, eds. *The Earthscan Reader on International Trade and Sustainable Development.* London, Earthscan.

Sun, X., Katsigris, E. & White, A. 2004. *China and forest trade in the Asia-Pacific region: implications for forests and livelihoods.* Bogor, Indonesia, Forest Trends.

Toyne, P., O'Brian, C. & Nelson, R. 2002. *The timber footprint of the G8 and China. Making the case for green procurement by government.* Gland, Switzerland, WWF International.

UNCTAD. 2003a. *Trade analysis and information system.* Geneva, Switzerland, United Nations Conference on Trade and Development.

UNCTAD. 2003b. *Back to basics: market access issues in the Doha Agenda.* Geneva, Switzerland.

UNECE/FAO. 2003. *Forest products annual market analysis 2002–2004*. Timber Bulletin LVI(3). Geneva, Switzerland, United Nations Economic Commission for Europe and FAO.

UNECE/FAO. 2004. *Forest products annual market review, 2003-2004*. Timber Bulletin LVII(3). Geneva, Switzerland (also available at www.unece.org/trade/timber/docs/fpama/2004/2004-fpamr.pdf).

WTO (World Trade Organization). 2001. *Doha WTO ministerial 2001: ministerial declaration*.

WT/MIN(01)/DEC/1, 20 November 2001. Geneva, Switzerland (available at www.wto.org/english/thewto_e/minist_e/min01_e/mindecl_e.htm).

WTO. 2003. *Understanding the WTO*. Geneva, Switzerland.

WTO. 2004. *Negotiations, implementation and development: the Doha agenda*. Geneva, Switzerland (available at www.wto.org/english/tratop_e/dda_e/dda_e.htm). ◆

Forests and war, forests and peace

Many violent conflicts occur in forested regions. This chapter outlines the reasons, and proposes some solutions. It notes the general characteristics of recent armed disputes, examines the links to and impact on forests, explores issues related to post-conflict situations and presents a strategy for action.

Although wars have been and are being fought all over the world, the focus in this chapter is on major clashes that have taken place in Africa, Latin America and South and Southeast Asia. Forest-related conflicts involving less violence exist in almost all countries. However, their characteristics and implications are somewhat different.

TRAGEDY OF VIOLENT CONFLICT
In 2000, 17 countries experienced armed conflict – defined as ongoing violence between two or more armies in which more than a thousand people die in combat – and another 12 countries were in post-conflict situations (Collier and Hoeffler, 2002). Although most of this unrest was caused by civil war, many of these wars affected neighbouring states as well. In addition, numerous countries experienced other forms of violence such as banditry, killings linked to land disputes and hostilities with fewer than a thousand fatalities.

The number of armed conflicts around the world rose steadily between 1965 and 1990, but has declined slightly since then. However, recent conflicts seem to last longer than in the past (Collier and Hoeffler, 2002).

Violent conflicts create huge economic and social costs, as well as a significant impact on the environment. At the end of a civil war, which on average lasts seven years, a country can expect its per capita income to be 15 percent lower and to have 30 percent more people living in poverty (Collier et al., 2003). In the past decade, millions of people, mostly civilians, have died in conflicts, and many more have been maimed or have had to flee. In 2001, the Office of the United Nations High Commissioner for Refugees (UNHCR) assisted 12 million refugees and 5.3 million internally displaced people (UNHCR, 2002). Indirect effects of conflict include a higher incidence of adult and child mortality, malaria and HIV / AIDS. In addition, most opium and cocaine production occurs in countries in conflict or post-conflict situations (Collier et. al., 2003).

GEOGRAPHY OF VIOLENT CONFLICT
While a complex set of political, ethnic, religious, ideological and economic factors accounts for why specific countries experience armed conflict, some are more prone to violence than others. Those most at risk have low per capita incomes and stagnant economies and export mostly primary commodities. About 50 countries, with a combined population of more than one billion, share all three characteristics. Countries where one ethnic group dominates a number of minorities are prime candidates for military hostilities, as are those that have suffered previous conflicts (Collier et al., 2003).

Although research on the geography of armed conflicts within national borders has not been widely conducted, studies (Goodhand, 2003; Le Billon, 2001; Starr, 2002) suggest that violent conflict is more frequent in areas that:
- are remote and inaccessible;
- have valuable natural resources in areas where property rights are uncertain or disputed;
- have a high proportion of poor households;
- have been poorly integrated into national democratic institutions;
- receive few public services;
- have several ethnic groups and religions.

Areas within countries that are most likely to experience armed conflict tend to be those with characteristics that provide the means or motives for war. They include secluded places where insurgents can hide and exploit valuable natural resources to finance military activities.

Inaccessibility and vegetative cover can also facilitate lucrative illegal activities such as cultivating illicit crops and smuggling. People may resort to violence to gain control over natural resources or because they feel neglected or mistreated. Often, motives are multiple and shift over time, combining political, religious or ethnic dimensions with personal incentives such as a desire for income, wealth, status, revenge or security or loyalty to specific individuals (Goodhand, 2003).

FORESTS AND VIOLENT CONFLICT
Forested regions in poor countries usually have many of the characteristics associated with locations where violent conflict occurs. They tend to be remote and inaccessible. They often have valuable timber, petroleum, land, ivory, diamonds, gold and other minerals, which insurgents can exploit or tax. Forest dwellers frequently resent that outsiders benefit most from these resources. Governments have tended to see forested regions as peripheral places with few people and little political importance or economic value, and have only focused on them to extract timber or minerals. Thus, forested regions have traditionally been poorly integrated into national political processes and receive few public services. Dominant ethnic groups have marginalized indigenous and tribal people in forested regions who also compete with migrants from other areas over resources. Given the limited employment opportunities in many forested regions, taking up arms can seem like an attractive way to earn a living.

The significance of this problem can be grasped by considering the following – incomplete – list of countries that have experienced armed conflicts during the past 20 years in forested areas: Angola, Bangladesh, Bosnia and Herzegovina, Cambodia, Central African Republic, Colombia, the Congo, Côte d'Ivoire, the Democratic Republic of the Congo, Guatemala, India, Indonesia, Liberia, Mexico, Mozambique, Myanmar, Nepal, Nicaragua, Pakistan, Peru, the Philippines, Rwanda, Senegal, Sierra Leone, Solomon Islands, Sri Lanka, the Sudan, Suriname and Uganda.

The forested regions of Bolivia, Brazil, the Lao People's Democratic Republic and Papua New Guinea have also suffered substantial social violence.

Forests as a means for war
Forests can provide refuge, funds and food for combatants. In many of the countries mentioned above, insurgents have used forested regions to hide from government troops. In the case of the Democratic Republic of the Congo, the lack of roads crossing the forested central and northern regions separated the eastern half of the country from the capital, making it much easier for insurgent groups and foreign armies to enter and remain there.

In several cases, governments ignored insurgents or made only perfunctory efforts to control them as long as they remained in remote forested regions, likely concluding that sustained military campaigns in areas of little strategic importance were too costly. Thus, insurgents in countries such as Colombia, Nepal and the Philippines were able gradually to build up military capacity.

Selling timber to fund armed activities is difficult because logs are easy to detect and operations require secure control over the territory. Nonetheless, such cases have been documented in Cambodia, the Democratic Republic of the Congo and Myanmar, and there may be others (Global Witness, 2003; Le Billon, 2000; UNSC, 2001). Insurgents mostly extract other natural resources found in forests to raise funds. For example, high-value metals such as columbo-tantalite (coltan) and cassiterite have been exploited in the Democratic Republic of the Congo in much the same way as diamonds and alluvial gold have in Angola, Liberia and Sierra Leone. These commodities, like ivory, do not require much capital and are easy to transport and conceal. Rebel groups have also been known to extort money from oil and mining companies and large farmers in remote areas. Similarly, armed groups or their supporters cultivate, sell and tax illicit crops grown in inaccessible forested mountain regions in Southeast and Central Asia and the foothills of the Andes.

Many governments use timber revenues to finance their armed forces, particularly in Central Africa and Southeast Asia. While sovereign states have an undeniable right to use their natural resources as they wish, sometimes national laws prohibit these activities, and officers use the proceeds for personal gain. It has also been alleged that, at times, military officials have prolonged conflicts so that they could engage in illegal logging and other illicit activities.

Forests and the motives for war

People rarely go to war over timber, but in Cambodia and Myanmar, for example, insurgent groups dragged out their struggles in part to engage in illegal logging operations (Le Billon, 2000; Global Witness, 2003). Similarly, while conflicts are seldom started to earn money from illicit crops in agricultural frontier areas, armed groups involved in the lucrative drug business have a strong incentive not to disarm. There are also cases in Africa and Asia where the desire to control oil and minerals in forested regions has resulted in conflicts lasting longer than they otherwise would have.

Forest-dwelling indigenous people and tribal groups have participated in violent conflicts in Bangladesh (Chittagong Hills), Guatemala (Quiché, Alta Verapaz), India (Assam, Nagaland), Indonesia (West Kalimantan and West Papua), the Lao People's Democratic Republic (north), Mexico (Chiapas), Myanmar (north) and Nicaragua (Atlantic Coast). Often marginalized or discriminated against, these people have reacted violently to outside attempts to curtail their activities, control their territories or exploit natural resources on their lands. Some have struggled for autonomy or independence, others for greater control over their natural resources, and still others for freedom to engage in their traditional activities or to cultivate illicit crops. Foreign governments and political ideologues have sometimes supported these movements and used them to promote their own political agendas. Their efforts find fertile ground because these groups feel politically disenfranchised. The Islamic

movements in the forested regions of Aceh in Indonesia and Mindanao in the Philippines offer examples of this trend. In parts of Africa, ethnic factors also shape conflicts in forested regions. This seems to be the case, for example, in the Democratic Republic of the Congo (North and South Kivu) and Senegal (Casamance).

In other instances, violent outbreaks are the result of long-standing government neglect or a weak and ineffective presence of central authorities. Such situations leave room for political activists and other groups to fill the void. Many Spanish-speaking agricultural frontier areas in tropical Latin America have witnessed widespread social violence, among them the Chapare in Bolivia, southeast Pará in Brazil, the Petén in Guatemala, the Peruvian Amazon and regions of Colombia. The rural insurgencies in Nepal and the Philippines have had similar characteristics.

Individuals often take up arms as the result of government troops violating human rights. This is prevalent in remote areas where there is less supervision and monitoring by the media and NGOs.

Some factors that favour violence in forested regions also apply to mountainous and arid districts, which can also be poor, isolated and neglected. Governments need to pay more attention to these historically marginalized areas, which are frequently inhabited by ethnic minorities.

IMPACT OF ARMED CONFLICT ON FORESTS

While war is almost always devastating for people, it has both negative and positive effects on forests (McNeely, 2003). Armies burn or clear forested areas with defoliants to spot the enemy more readily, and soldiers hunt wildlife for food (Hart and Mwinyihali, 2001; SAMFU, 2002). Land mines not only kill and maim people, but also kill and maim gorillas and other large mammals. When access to remote forests is cut off, forests in secure locations become threatened. Loggers and farmers often take advantage of roads built for military purposes to exploit resources that run through forested areas.

Large concentrations of refugees and displaced people frequently put great pressure on their local environment, as occurred in Rwanda

mines over large tracts tends to keep farmers and other civilians out, thereby promoting conservation and natural regeneration in places that have recently been cleared.

War also discourages investment in logging and in destroying forests to plant pastures and tree crops. Ranchers concerned about being kidnapped or having their cattle stolen tend to stay away, and logging companies may be unwilling to risk losing valuable machinery. The decrease in investment in such activities is usually harmful for the economy and, in some cases, for the long-term management of the forest. In the short term, however, it protects the resource.

POST-CONFLICT

Post-conflict situations also pose specific challenges. First, 44 percent of countries affected by conflict return to war within five years of a ceasefire (Collier *et al.*, 2003). After the conflict ends, many of the factors that caused it are often still present, and some may even have been aggravated. Such circumstances undermine economic growth and generate more unemployment, especially among youth. Moreover, former combatants and their political supporters often find it difficult to reintegrate into civilian life. Thus, they may be easily persuaded to take up arms again to earn a living and regain their former status.

Following a war, governments and international aid agencies are too preoccupied with other concerns to focus on longer-term issues such as forest management or conservation. They are under tremendous pressure to restore the economy, and logging is often the only option in many low-income countries. Harvesting activities therefore frequently expand much more rapidly than the public sector's capacity to regulate them, as was the case in Cambodia and is likely to be in the Democratic Republic of the Congo and Liberia. Post-war governments in Colombia, Guatemala and Nicaragua relied on forested areas to settle demobilized soldiers and displaced people, as those were the only large areas of sparsely populated lands (Kaimowitz, 2002).

Conflicts also tend to strengthen the power of the military, making it difficult for civilian authorities to render army personnel accountable for their actions. Some governments have encouraged their forces to engage in economic activities such as logging rather than fund operations from the central budget. In countries where the military is involved in logging or is closely associated with private logging companies, or where the government allows private forestry companies to establish their own militia for protection purposes, it is much more difficult to enforce forestry and conservation laws (Carle, 1998).

Large concentrations of refugees and displaced people frequently put great pressure on their local environment (Hart and Mwinyihali, 2001; Plumptre, 2003). They move into new areas to hunt, fish, collect fuelwood and cut trees to build houses. Such actions can rapidly deplete local resources.

Moreover, war drains away funds that governments might otherwise invest in forestry activities, and civil servants may be helpless to act when military officials or armed insurgents engage in predatory logging (Plumptre, 2003).

Paradoxically, war can also be beneficial for forests (McNeely, 2003). Conflict forces large numbers of families to flee rural areas, allowing regeneration to take place in the abandoned areas (Alvarez, 2003). Similarly, the presence of land

After conflict, farmers, ranchers and loggers return to resume their activities in rural areas. In addition, heavily armed unemployed young men, with few choices besides logging, commercial hunting and banditry, often join them. Governments may promise former combatants land, training and credit but be unable to keep or sustain these commitments over time. Their inability to deliver sows the seeds for future conflict.

FORESTS FOR PEACE – A STRATEGY FOR ACTION

Efforts to promote peace in forested regions must start with removing the motives for conflict before it breaks out. Governments need to take bold steps to recognize the political, cultural and territorial rights of ethnic minorities and others living in forested regions. They need to integrate forest-dependent people into the wider economy and national political life without marginalizing them or forcing them to abandon their homes and cultures. They also need to provide social services and greater access to markets without encouraging an influx of outside settlers (Goodhand, 2003). Small-scale agricultural, forestry, fishing and handicraft projects can open up new livelihood options and reduce the vulnerability of people living in forested regions. Sustainable industrial timber harvesting would also help, as would a more equitable sharing of the benefits from forest resources. A well-trained police force, a judicial system that respects local practices and a system of independent monitoring of human rights violations are other key elements that would increase the feeling of security of local populations.

Such measures, although costly, can be justified in terms of the ethical imperative to promote social justice and sustainable development. However, since resources at a country's disposal are usually limited, investment in these types of expenditure often occurs in more accessible and densely populated areas. Governments need to appreciate that per capita investments required in forested regions, while high, are much lower than the cost of armed conflict, once it breaks out.

Where war does break out, forests and environmental concerns can be important in the peace process. In various countries, both government forces and insurgents have agreed to respect certain environmental rules, recognizing that doing so is in the best interests of the population. In Rwanda, for example, an agreement was reached to avoid harming mountain gorillas during the civil war because of their importance to the tourism industry (Plumptre, 2003). Maoist insurgents in Nepal largely respected the government's community forestry programme, reflecting the widespread legitimacy it had gained. In addition, during peace negotiations between the government and the Revolutionary Armed Forces of Colombia (FARC) in 2000, both sides included environmental issues in the first rounds of talks (Alvarez, 2003).

Sanctions to stop timber sales by insurgent groups and by de facto governments that are not recognized by the international community can sometimes be effective, as can efforts to control money laundering associated with such activities. UN agencies have imposed such sanctions at different times in Cambodia and Liberia (UNSC, 2003) and suggested similar action be taken in the Democratic Republic of the Congo. In the first case, implementation was uneven and the results mixed, but sanctions ultimately played an important role in the collapse of the Khmer Rouge. In Liberia, the war ended before the effectiveness of the sanctions, enforced in response to interference by the government in the internal affairs of its neighbours, could be fully assessed.

It is important to include forestry and other natural resource issues on the agenda in peace negotiations because of their economic importance to opposing factions. Involving separatist movements in controlling timber and other natural resources, with the tax revenues they provide, is key to defining viable options for regional autonomy. Such considerations have been the subject of lengthy discussions concerning the Atlantic Coast of Nicaragua; Aceh and West Papua, Indonesia; Mindanao,

Forests, wood and war in European history

Until recently, access to sufficient supplies of wood was a vital part of strategic readiness in European countries. Uses included fuel, housing, wagons and temporary applications such as supporting trenches in the First World War. Some purposes required wood with specific characteristics, notably in ship building, which demanded softwood for masts and hardwood for keels. Such specialized requirements stimulated exports from Nordic countries or, in the case of ancient Greece, exports from the Black Sea region to Attica. Along similar lines, one of England's motives in colonizing Canada was to ensure access to high-quality wood for masts for the Royal Navy. The incentive for France's adoption of modern silviculture in the seventeenth century was to secure a domestic source of oak for the French navy. Strategic reasons were also cited for the United Kingdom's afforestation programme in the 1920s.

Forest resources were overused, for example, to supply timber for military ships and energy for armies or displaced populations. During and immediately after both world wars, European forests were systematically overcut and principles of sustainable yield forgotten. In the Second World War, one of Germany's strategic weaknesses was lack of access to oil reserves, a weakness it tried to minimize by developing wood chemistry as a replacement fuel. Plans were made to use millions of hectares of plantations in Eastern Europe to supply wood feedstock for this new chemical industry.

Even today, the wood from some forests in eastern France is of lower value because of the

bullets and shrapnel embedded in it as a result of battles in the First World War. Among other hazards, harvesting causes injuries and breaks costly equipment. In other armed conflicts, orchards and olive groves were deliberately destroyed as an economic or symbolic act.

The collapse of institutions, authority and morality that tends to occur during some armed conflicts and most civil wars removes an important protection from forests. During the war in former Yugoslavia, it was said that several warlords built large fortunes as a result of illegal logging and export of wood from the country's high-quality forests. When Sarajevo was under siege during the conflict in Bosnia and Herzegovina, the population cut down all forests on the hills surrounding the city, up to the line of the Serb occupation, because they needed fuelwood to get through the winter.

In a few extreme cases, war has helped forests by allowing ecosystems to recover free of human pressure. For example, in the Hundred Years' War between England and France in the fourteenth century, the destruction and subsequent abandonment of many villages had a positive effect on forests.

the Philippines; the Jaffna peninsula, Sri Lanka; and various regions in Myanmar, for example.

Conservation organizations working in conflict situations must take care to maintain their neutrality. They should avoid taking steps that create resentment by curtailing local people's access to natural resources and encourage all sides to recognize the benefits of conservation. It is crucial to take appropriate security precautions

and to rely heavily on local personnel who understand the context and the terrain (Hart and Mwinyihali, 2001; Shambaugh *et al.*, 2001).

Refugee relief and humanitarian organizations need to strengthen their commitment to prevent refugee camps from destroying the environment. Guidelines approved by UNHCR in 1996 mark a major step forward in this regard but have yet to be consistently put into practice (UNHCR, 1996).

Once hostilities cease, the international community can offer long-term and more generous assistance to help restore forest resources, as in this afforestation project in Viet Nam

Ideally, post-war environmental and forestry planning should begin while the conflict is still in progress. Discussions should address where to relocate demobilized troops and the support they will receive to reintegrate into civilian life. It is also essential to bring stakeholders together to discuss how to deal with forests and the environment during the transition period. Because governments are likely to concentrate their resources on the war, the onus on international donors to finance most of these efforts is generally significant.

Once hostilities cease, the international community can help improve conditions by offering long-term and more generous assistance, given that countries have an urgent need for foreign exchange. Such external assistance may prevent the exploitation of forest resources before a suitable regulatory framework is put in place. Since most government agencies in post-conflict situations are weak, new rules need to be simple and focused on a few key activities. Independent monitoring is also critically important.

In post-war situations, countries should not wait to address issues related to the management of natural resources and the environment until peace and economic growth are restored. Cambodia, Liberia, Myanmar and Solomon Islands, for example, have few sources of foreign exchange besides forest products. Therefore,

sustainable production and more equitable sharing of benefits must figure prominently in economic development to keep them from falling back into violent conflict. The same applies, to a lesser extent, to Angola, the Congo, the Democratic Republic of the Congo and Sierra Leone, where petroleum and minerals dominate exports and timber is secondary.

Parks located along sensitive borders can help reduce tensions and foster cooperation between neighbouring countries that have traditionally distrusted each other. The Condor Peace Park along the disputed border between Ecuador and Peru is one example. An added advantage is that such parks can attract financial resources and increase non-military presence. In some cases, there may be opportunities to employ former combatants to plant and protect trees.

In summary, forested regions in some countries provide the motives for engaging in war and the capacity to do so. Thus, they are prone to armed hostilities, which can have both negative and positive effects. However, post-conflict situations in countries with significant forests almost always pose an acute danger for this resource. The international community, national governments and others need to make special efforts to avoid the outbreak of conflict in such areas, use forest-related measures to resolve conflicts when they do occur, reduce the environmental impact of conflict and use forests to promote peace and prosperity in post-conflict situations.

Peace requires a commitment that begins with investing in better governance and improving livelihoods in remote forested and mountainous regions. If these areas are prevented from serving as breeding grounds for violence, forests can assume their rightful importance for the social, cultural, economic and environmental contributions they make to the lives of all who depend on them. ◆

REFERENCES

Alvarez, M.D. 2003. Forests in the time of violence: conservation implications of the Colombian war. *Journal of Sustainable Forestry*, 16(3–4): 49–70.

Carle, J. 1998. *Forest industries and log trade policy in Cambodia*. Technical Paper No. 4, Forest Policy Reform Project. Burlington, USA, Associates in Rural Development.

Collier, P. & Hoeffler, A. 2002. *Greed and grievance in civil war*. CSAE Working Paper Series No. 2002-01. Oxford, UK, Centre for the Study of African Economies (available at www.csae.ox.ac.uk).

Collier, P., Elliott V., Hegre, H., Hoeffler, A., Reyna-Querol, M. & Sambinas, N. 2003. *Breaking the conflict trap, civil war and development policy*. Washington, DC, World Bank.

Global Witness. 2003. *A conflict of interests, the uncertain future of Burma's forests*. London, Global Witness.

Goodhand, J. 2003. Enduring disorder and persistent poverty: a review of the linkages between war and chronic poverty. *World Development*, 31(3): 629–646.

Hart, T. & Mwinyihali, R. 2001. *Armed conflict and biodiversity in sub-Saharan Africa: the case of the Democratic Republic of Congo (DRC)*. Washington, DC, Biodiversity Support Program.

Kaimowitz, D. 2002. Resources, abundance and competition in the Bosawas Biosphere Reserve, Nicaragua. *In* R. Matthew, M. Halle & J. Switzer, eds. *Conserving the peace: resources, livelihoods and security*, pp. 171–198. Winnipeg, Canada, International Institute for Sustainable Development.

Le Billon, P. 2000. The political ecology of transition in Cambodia, 1989–1999: war, peace and forest exploitation. *Development and Change*, 31(4): 785–805.

Le Billon, P. 2001. The political ecology of war: natural resources and armed conflicts. *Political Geography*, 20: 561–584.

McNeely, J. 2003. Biodiversity, war and tropical forests. *Journal of Sustainable Forestry*, 16(3–4): 1–20.

Plumptre, A.J. 2003. Lessons learned from on-the-ground conservation in Rwanda and the Democratic Republic of the Congo. *Journal of Sustainable Forestry*, 16(3–4): 71–92.

SAMFU. 2002. *Plunder: the silent destruction of Liberia's rainforest*. Monrovia, Save My Future Foundation (available at www.forestsmonitor.org/reports/plunder/plunder.pdf).

Shambaugh, J., Ogelthorpe, J., Ham, R. & Tognetti, S. 2001. *The trampled grass: mitigating the impacts of armed conflict on the environment*. Washington, DC, Biodiversity Support Program.

Starr, S.F. 2002. *Conflict and peace in mountain societies*. Thematic Paper for Bishkek Global Mountain Summit. Nairobi, United Nations Environment Programme.

UNHCR. 1996. *UNHCR environmental guidelines*. Geneva, Switzerland, Office of the United Nations High Commissioner for Refugees (available at www.unhcr.ch).

UNHCR. 2002. *Refugees by numbers*. Geneva, Switzerland (available at www.unhcr.ch).

UNSC. 2001. *Addendum to the report of the panel of experts on the illegal exploitation of natural resources and other forms of wealth of the Democratic Republic of the Congo*. S/2001/1072 (13 November 2001). New York, USA, United Nations Security Council (available at www.un.org/Docs/sc/letters/2001/sglet01.htm).

UNSC. 2003. *Resolution 1478 (2003)*. S/RES/1478 (6 May 2003). New York, USA (available at www.un.org/Docs/sc/unsc_resolutions03.html). ◆

ANNEX 1
ACRONYMS

ACRONYMS

AFWC
African Forestry and Wildlife Commission

AOSIS
Alliance of Small Island States

APFC
Asia-Pacific Forestry Commission

ATO
African Timber Organization

CATIE
Tropical Agricultural Research and Higher Education Center

CBD
Convention on Biological Diversity

CDM
Clean Development Mechanism (Kyoto Protocol)

CICI 2003
International Conference on the Contribution of Criteria and Indicators for Sustainable Forest Management: the Way Forward

CIFOR
Center for International Forestry Research

CITES
Convention on International Trade in Endangered Species of Wild Fauna and Flora

COFO
Committee on Forestry (FAO)

CO$_2$
carbon dioxide

COP
Conference of the Parties

CPF
Collaborative Partnership on Forests

CTE
Committee on Trade and Environment (WTO)

ECOSOC
Economic and Social Council (UN)

EFC
European Forestry Commission

EU
European Union

FARC
Revolutionary Armed Forces of Colombia

FLEG
Forest Law Enforcement and Governance

FLEGT
Forest Law Enforcement, Governance and Trade

FORSPA
Forestry Research Support Programme for Asia and the Pacific

FRA
Global Forest Resources Assessment

GDP
gross domestic product

GEF
Global Environment Facility

GFIS
Global Forest Information Service

GFMC
Global Fire Monitoring Center

GM
genetic modification

GMO
genetically modified organism

GNP
gross national product

GPG
Good Practice Guidance for Land Use, Land-Use Change and Forestry

GSP
generalized system of preferences

HS
Harmonized System (WCO)

ICRAF
World Agroforestry Centre (formerly International
Council for Research in Agroforestry)

ICS
Incident Command System

IFF
Intergovernmental Forum on Forests

IFIA
Interafrican Forest Industries Association

ILO
International Labour Organization

IMFNS
International Model Forest Network Secretariat

INBAR
International Network for Bamboo and Rattan

IPCC
Intergovernmental Panel on Climate Change

IPF
Ad Hoc Intergovernmental Panel on Forests

ISIC
International Standard Industrial Classification
of all Economic Activities (ILO)

ITTA
International Tropical Timber Agreement

ITTO
International Tropical Timber Organization

IUCN
World Conservation Union

IUFRO
International Union of Forest Research Organizations

MAB
Man and the Biosphere (UNESCO)

MCPFE
Ministerial Conference on the Protection of Forests
in Europe

NEFC
Near East Forestry Commission

NEPAD
New Partnership for Africa's Development

NFAP
National Forestry Action Programme
(previously National Forestry Action Plan)

NGO
non-governmental organization

NWFP
non-wood forest product

OECD
Organisation for Economic Co-operation and
Development

PROFOR
Program on Forests

SADC
Southern African Development Community

SEEA
System of Integrated Environmental
and Economic Accounting

SIDS
small island developing states

SPS
Sanitary and Phytosanitary Measures (WTO Agreement)

TBT
Technical Barriers to Trade (WTO Agreement)

TFAP
Tropical Forestry Action Programme
(previously Tropical Forestry Action Plan)

TIMO
timberland investment management organization

TPES
total primary energy supply

UN
United Nations

UNCCD
United Nations Convention to Combat Desertification

UNCED
United Nations Conference on Environment and Development

UNCTAD
United Nations Conference on Trade and Development

UNDP
United Nations Development Programme

UNECE
United Nations Economic Commission for Europe

UNEP
United Nations Environment Programme

UNEP-WCMC
UNEP World Conservation Monitoring Centre

UNESCO
United Nations Educational, Scientific and Cultural Organization

UNFCCC
United Nations Framework Convention on Climate Change

UNFF
United Nations Forum on Forests

UNHCR
(Office of the) United Nations High Commissioner for Refugees

WCO
World Customs Organization

WRI
World Resources Institute

WWF
World Wide Fund for Nature

WSSD
World Summit on Sustainable Development

WTO
World Trade Organization

ANNEX 2
DATA TABLES

EXPLANATORY NOTES

GENERAL

Country/area nomenclature and regional groups used in the data tables

The country/area names and order used in these tables follow standard UN practice regarding nomenclature and alphabetical listing. Data for "China" incorporate values for China (including Hong Kong Special Administrative Region and Macao Special Administrative Region) and for Taiwan Province of China. The regional groups used in these tables represent FAO's standardized regional breakdown of the world according to geographic – not economic or political – criteria.

Totals

Regional and global totals may not tally because of rounding or territories not included in the tables.

Abbreviations

n.s. = not significant, indicating a very small value

- = not available

TABLE 1

"Land area" refers to the total area of a country, excluding areas under inland water bodies. The source of these data is FAO (2001).

All population figures are from FAO (2004).

Economic data are from World Bank (2004). The gross domestic product (GDP) per capita figure represents the GDP divided by the mid-year population. The data are in constant 1995 US dollars. The annual percentage growth rate of GDP is based on constant local currency.

TABLES 2 AND 3

These figures for 2000 represent the most current global data set available for forest area and forest area change. The source of the data is FAO (2001). In Table 2, "total forest" is the sum of natural forest plus forest plantations. Forest area change is the net change in forests and includes expansion of forest plantations and losses and gains in the area of natural forests.

In Table 3, "volume" refers to total volume over bark of living trees above 10 cm diameter at breast height. "Biomass" refers to above-ground mass of the woody part (stem, bark, branches, twigs) of trees (alive or dead), shrubs and bushes. For Europe, the countries of the Commonwealth of Independent States, Australia, Canada, Japan, New Zealand and the United States, the stem volume for all living trees has been used for the volume figure. Some variation as to the minimum diameter applied is reported in ECE/FAO (2000).

TABLE 4

The source of the data is FAO (2004).

"0" indicates either a true zero or an insignificant value (less than half a unit).

TABLE 5

The source of information is untreaty.un.org/ENGLISH/bible/englishinternetbible/partI/chapterXXVII/chapterXXVII.asp as well as the Web sites of the listed conventions and agreements:

- CBD: www.biodiv.org/world/parties.asp
- UNFCCC: unfccc.int/resource/conv/ratlist.pdf
- Kyoto Protocol: unfccc.int/resource/kpstats.pdf
- CCD: www.unccd.int/convention/ratif/doeif.php?sortby=name
- CITES: www.cites.org/eng/disc/parties/index.shtml
- Ramsar Convention: www.ramsar.org/key_cp_e.htm
- World Heritage Convention: whc.unesco.org/nwhc/pages/doc/main.htm

In addition to the countries indicated in the table, the European Community has ratified CBD, UNFCCC, the Kyoto Protocol and CCD.

REFERENCES

FAO. 2001. *Global Forest Resources Assessment 2000 – main report.* FAO Forestry Paper No. 140. Rome (available at www.fao.org/forestry/fo/fra/main/index.jsp).

FAO. 2004. *FAOSTAT statistical database.* Rome (available at apps.fao.org/faostat/collections).

United Nations Economic Commission for Europe (ECE)/FAO. 2000. *Forest resources of Europe, CIS, North America, Australia, Japan and New Zealand (industrialized temperate/boreal countries).* New York, USA, and Geneva, Switzerland, UN.

World Bank. 2004. *World Development Indicators 2004.* Washington, DC.

TABLE 1
Basic data on countries and areas

Country/area	Total land area ('000 ha)	Population				Economic indicators	
		Total, 2003 ('000)	Density, 2003 (population/km²)	Annual rate of change, 2000–2005 (%)	Rural, 2003 (%)	GDP per capita, 2003 (US$)	Annual growth rate of GDP, 2002 (%)
Africa, total	**2 978 394**	**850 558**	**28.6**	**2.2**	**61.3**	...	...
Algeria	238 174	31 800	13.4	1.7	41.2	2 049	4.1
Angola	124 670	13 625	10.9	3.1	64.3	725	15.3
Benin	11 063	6 736	60.9	2.6	55.4	521	6.0
Botswana	56 673	1 785	3.1	0.8	48.4	3 983	3.1
Burkina Faso	27 360	13 002	47.5	2.9	82.2	294	4.6
Burundi	2 568	6 825	265.8	3.1	90.1	86	3.6
Cameroon	46 540	16 018	34.4	1.8	48.6	803	4.4
Cape Verde	403	463	114.9	1.9	44.1	1 765	4.6
Central African Republic	62 297	3 865	6.2	1.3	57.3	325	-0.8
Chad	125 920	8 598	6.8	2.9	75.1	290	9.9
Comoros	186	768	412.9	2.7	65.0	369	3.0
Congo	34 150	3 724	10.9	2.6	46.5	1 050	3.5
Cote d'Ivoire	31 800	16 631	52.3	1.6	55.1	886	-1.8
Dem. Republic of the Congo	226 705	52 771	23.3	2.8	68.4	107	3.0
Djibouti	2 317	703	30.3	1.6	16.3	886	1.6
Egypt	99 545	71 931	72.3	2.0	57.9	1 062	3.0
Equatorial Guinea	2 805	494	17.6	2.6	51.9	5 915	16.2
Eritrea	11 759	4 141	35.2	3.6	80.1	305	1.8
Ethiopia	110 430	70 678	64.0	2.4	84.4	91	2.7
Gabon	25 767	1 329	5.2	1.8	16.2	4 155	3.0
Gambia	1 000	1 426	142.6	2.6	73.9	224	-3.1
Ghana	22 754	20 922	91.9	2.1	54.6	354	4.5
Guinea	24 572	8 480	34.5	1.6	65.1	424	4.2
Guinea-Bissau	3 612	1 493	41.3	2.9	66.0	208	-7.2
Kenya	56 915	31 987	56.2	1.4	60.6	444	1.0
Lesotho	3 035	1 802	59.4	0.2	82.1	594	3.8
Liberia	11 137	3 367	30.2	3.9	53.3	181	3.3
Libyan Arab Jamahiriya	175 954	5 551	3.2	1.9	13.7	3 640	-
Madagascar	58 154	17 404	29.9	2.8	73.5	318	-12.7
Malawi	9 409	12 105	128.7	2.0	83.7	158	1.8
Mali	122 019	13 007	10.7	3.0	67.7	298	4.4
Mauritania	102 522	2 893	2.8	2.9	38.2	381	3.3
Mauritius	202	1 221	604.5	1.0	56.7	4 594	4.4
Morocco	44 630	30 566	68.5	1.6	42.5	1 463	3.2
Mozambique	78 409	18 863	24.1	1.7	64.4	222	7.7
Namibia	82 329	1 987	2.4	1.4	67.6	2 307	2.7
Niger	126 670	11 972	9.5	3.6	77.8	227	3.0
Nigeria	91 077	124 009	136.2	2.5	53.3	390	-0.9
Réunion	250	756	302.4	1.5	8.5	14 614	-
Rwanda	2 466	8 387	340.1	2.1	81.7	185	9.4
Saint Helena	31	5	16.1	-	64.5	-	-
Sao Tome and Principe	95	161	169.5	2.5	62.4	364	4.1
Senegal	19 252	10 095	52.4	2.4	50.4	641	1.1
Seychelles	45	81	180.0	-	50.1	8 814	0.3
Sierra Leone	7 162	4 971	69.4	3.7	61.2	197	6.3
Somalia	62 734	9 890	15.8	4.1	65.2	155	-
South Africa	121 758	45 026	37.0	0.6	43.1	3 551	3.0
Sudan	237 600	33 610	14.1	2.1	61.1	459	5.5
Swaziland	1 721	1 077	62.6	0.8	76.5	1 653	3.6
Togo	5 439	4 909	90.3	2.3	64.9	377	4.6
Tunisia	16 362	9 832	60.1	1.1	36.3	2 561	1.7
Uganda	19 964	25 827	129.4	3.2	87.8	242	6.7
United Republic of Tanzania	88 359	36 977	41.8	1.9	64.6	271	6.3
Western Sahara	26 600	308	1.2	2.6	6.3	-	-
Zambia	74 339	10 812	14.5	1.2	64.3	398	3.3
Zimbabwe	38 685	12 891	33.3	0.5	65.1	190	-5.6

Note: The regional breakdown reflects geographic rather than economic or political groupings.

Country/area	Total land area ('000 ha)	Population				Economic indicators	
		Total, 2003 ('000)	Density, 2003 (population/km²)	Annual rate of change, 2000–2005 (%)	Rural, 2003 (%)	GDP per capita, 2003 (US$)	Annual growth rate of GDP, 2002 (%)
Asia, total	**3 084 746**	**3 823 390**	**123.9**	**1.2**	**61.2**	**...**	**...**
Afghanistan	64 958	23 897	36.8	3.8	76.7	167	-
Armenia	2 820	3 061	108.5	-0.5	35.6	905	12.9
Azerbaijan	8 359	8 370	100.1	0.9	50.0	853	10.6
Bahrain	69	724	1 049.3	2.1	10.0	12 542	3.5
Bangladesh	13 017	146 736	1 127.3	2.0	75.8	385	4.4
Bhutan	4 701	2 257	48.0	2.9	91.5	303	7.7
Brunei Darussalam	527	358	67.9	2.2	23.8	12 919	-
Cambodia	17 652	14 144	80.1	2.4	81.4	278	5.5
China	932 743	1 304 196	139.8	0.7	61.4	1 100	8.0
Cyprus	925	802	86.7	0.7	30.8	16 038	2.0
Dem. People's Rep. of Korea	12 041	22 664	188.2	0.5	38.9	494	-
Georgia	6 831	5 126	75.0	-0.9	48.1	770	5.6
India	297 319	1 065 462	358.4	1.5	71.7	555	4.6
Indonesia	181 157	219 883	121.4	1.3	54.4	944	3.7
Iran, Islamic Rep. of	162 201	68 920	42.5	1.2	33.3	2 079	6.7
Iraq	43 737	25 175	57.6	2.6	32.8	594	-
Israel	2 062	6 433	312.0	2.0	8.4	18 101	-0.8
Japan	37 652	127 654	339.0	0.1	34.6	33 819	0.3
Jordan	8 893	5 473	61.5	2.6	21.0	1 803	4.9
Kazakhstan	267 074	15 433	5.8	-0.4	44.2	1 785	9.8
Kuwait	1 782	2 521	141.5	5.1	3.7	13 641	-1.0
Kyrgyzstan	19 180	5 138	26.8	1.1	66.1	372	-0.5
Lao People's Dem. Rep	23 080	5 657	24.5	8.2	79.3	361	5.0
Lebanon	1 024	3 653	356.7	0.2	12.5	5 023	1.0
Malaysia	32 855	24 425	74.3	0.1	36.1	4 227	4.1
Maldives	30	318	1 060.0	198.1	71.2	2 260	5.6
Mongolia	156 650	2 594	1.7	21.4	43.3	462	4.0
Myanmar	65 755	49 485	75.3	0.2	70.6	1 174	-
Nepal	14 300	25 164	176.0	14.7	85.0	233	-0.5
Oman	21 246	2 851	13.4	49.8	22.4	7 389	0.0
Pakistan	77 087	153 578	199.2	0.0	65.9	498	2.8
Palestine, Occupied Territory	618	2 737	433	-	-	-	-
Philippines	29 817	79 999	268.3	0.3	39.0	1 005	4.4
Qatar	1 100	610	55.5	114.1	8.0	34 685	-
Republic of Korea	9 873	47 700	483.1	0.1	19.7	11 059	6.3
Saudi Arabia	214 969	24 217	11.3	0.6	12.3	8 561	1.0
Singapore	61	4 253	6 972.1	9.8	0.0	21 195	2.2
Sri Lanka	6 463	19 065	295.0	0.3	79.0	913	4.0
Syrian Arab Republic	18 377	17 800	96.9	3.5	49.9	1 497	2.7
Tajikistan	14 087	6 245	44.3	16.1	75.3	249	9.1
Thailand	51 089	62 833	123.0	0.1	68.1	2 273	5.4
Timor-Leste	1 479	871	58.9	3.6	92.4	434	-0.5
Turkey	76 963	71 325	92.7	0.1	33.7	3 418	7.8
Turkmenistan	46 992	4 867	10.4	8.0	54.7	3 078	14.9
United Arab Emirates	8 360	2 995	35.8	36.4	14.9	22 130	1.8
Uzbekistan	41 424	26 093	63.0	-0.9	63.4	338	4.2
Viet Nam	32 550	81 377	250.0	1.5	74.3	471	7.0
Yemen	52 797	20 010	37.9	3.5	74.4	484	3.6
Europe, total	**2 259 957**	**726 338**	**32.1**	**-0.1**	**27.0**	**...**	**...**
Albania	2 740	3 166	115.5	0.7	56.2	1 915	4.7
Andorra	45	71	157.8	4.0	8.3	20 424	-
Austria	8 273	8 116	98.1	0.0	34.2	31 187	1.0
Belarus	20 748	9 895	47.7	-0.5	29.1	1 768	4.7
Belgium and Luxembourg	3 282	10 318	314.4	0.2	2.8	29 257	0.7
Bosnia and Herzegovina	5 100	4 161	81.6	1.1	55.7	1 613	3.9
Bulgaria	11 055	7 897	71.4	-0.8	30.2	2 533	4.8
Croatia	5 592	4 428	79.2	-0.2	41.0	6 398	5.2
Czech Republic	7 728	10 236	132.5	-0.1	25.7	8 834	2.0
Denmark	4 243	5 364	126.4	0.2	14.7	39 497	2.1

Note: The regional breakdown reflects geographic rather than economic or political groupings.

Country/area	Total land area ('000 ha)	Population				Economic indicators	
		Total, 2003 ('000)	Density, 2003 (population/km²)	Annual rate of change, 2000–2005 (%)	Rural, 2003 (%)	GDP per capita, 2003 (US$)	Annual growth rate of GDP, 2002 (%)
Estonia	4 227	1 323	31.3	-1.1	30.6	6 232	6.0
Finland	30 459	5 207	17.1	0.2	39.1	31 069	1.6
France	55 010	60 144	109.3	0.5	23.7	29 222	1.2
Germany	34 927	82 476	236.1	0.1	11.9	29 137	0.2
Greece	12 890	10 976	85.2	0.1	39.2	15 690	4.0
Hungary	9 234	9 877	107.0	-0.5	34.9	8 384	3.3
Iceland	10 025	290	2.9	0.7	7.2	36 328	-0.5
Ireland	6 889	3 956	57.4	1.1	40.1	38 864	6.9
Italy	29 406	57 423	195.3	-0.1	32.6	25 527	0.4
Latvia	6 205	2 307	37.2	-1.0	33.8	4 453	6.1
Liechtenstein	15	34	226.7	1.3	78.4	43 486	-
Lithuania	6 258	3 444	55.0	-0.6	33.3	5 203	6.7
Malta	32	394	1 231.3	0.5	8.3	11 790	1.5
Netherlands	3 392	16 149	476.1	0.5	34.2	31 759	0.2
Norway	30 683	4 533	14.8	0.4	21.4	48 881	1.0
Poland	30 442	38 587	126.8	-0.1	38.1	5 355	1.4
Portugal	9 150	10 062	110.0	0.1	45.4	14 645	0.4
Republic of Moldova	3 296	4 267	129.5	-0.1	54.0	459	7.2
Romania	23 034	22 334	97.0	-0.2	45.5	2 550	4.3
Russian Federation	1 688 851	143 246	8.5	-0.6	26.7	3 026	4.3
San Marino	6	28	466.7	1.3	3.8	38 397	-
Serbia and Montenegro	10 200	10 527	103.2	-0.1	48.0	1 843	4.0
Slovakia	4 808	5 402	112.4	0.1	42.6	6 019	4.4
Slovenia	2 012	1 984	98.6	-0.1	49.2	13 831	2.9
Spain	49 945	41 060	82.2	0.2	23.5	20 424	2.0
Sweden	41 162	8 876	21.6	0.1	16.6	33 925	1.9
Switzerland	3 955	7 169	181.3	0.0	32.5	43 486	0.1
The FYR of Macedonia	2 543	2 056	80.8	0.5	40.5	2 225	0.7
Ukraine	57 935	48 523	83.8	-0.8	32.8	975	4.8
United Kingdom	24 160	59 251	245.2	0.3	10.9	30 355	1.8
North and Central America, total	**2 136 966**	**506 534**	**23.7**	**1.6**	**24.1**	**...**	**...**
Antigua and Barbuda	44	73	165.9	0.5	62.3	9 036	2.9
Bahamas	1 001	314	31.4	1.3	10.5	14 462	-
Barbados	43	270	627.9	0.4	48.3	9 867	-2.1
Belize	2 280	256	11.2	2.0	51.7	3 363	3.7
Bermuda	5	82	1 640.0	0.8	0.0	51 991	-
British Virgin Islands	15	21	140.0	2.7	38.9	-	-
Canada	922 097	31 510	3.4	0.8	19.6	27 097	3.3
Cayman Islands	26	40	153.8	3.7	0.0	-	-
Costa Rica	5 106	4 173	81.7	1.9	39.4	4 189	3.0
Cuba	10 982	11 300	102.9	0.3	24.4	2 762	1.1
Dominica	75	79	105.3	-0.1	28.0	3 279	-5.2
Dominican Republic	4 838	8 745	180.8	1.5	40.7	2 408	4.1
El Salvador	2 072	6 515	314.4	1.5	40.4	2 302	2.1
Greenland	34 170	57	0.2	0.1	17.6	-	-
Grenada	34	80	235.3	0.3	59.3	4 262	1.2
Guadeloupe	169	440	260.4	0.9	0.3	14 518	-
Guatemala	10 843	12 347	113.9	2.5	53.7	1 963	2.2
Haiti	2 756	8 326	302.1	1.3	62.5	300	-0.9
Honduras	11 189	6 941	62.0	2.3	54.4	980	2.5
Jamaica	1 083	2 651	244.8	0.9	47.9	2 802	1.1
Martinique	107	393	367.3	0.5	4.3	14 504	-
Mexico	190 869	103 457	54.2	1.4	24.5	5 945	0.9
Montserrat	11	4	36.4	-0.3	86.6	-	-
Netherlands Antilles	80	221	276.3	0.9	30.3	13 261	-
Nicaragua	12 140	5 466	45.0	2.4	42.7	750	1.0
Panama	7 443	3 120	41.9	1.8	42.9	3 400	0.8
Puerto Rico	887	3 879	437.3	0.5	3.3	20 812	-
Saint Kitts and Nevis	36	42	116.7	-0.8	67.8	8 927	2.1

Note: The regional breakdown reflects geographic rather than economic or political groupings.

Country/area	Total land area ('000 ha)	Population				Economic indicators	
		Total, 2003 ('000)	Density, 2003 (population/km²)	Annual rate of change, 2000–2005 (%)	Rural, 2003 (%)	GDP per capita, 2003 (US$)	Annual growth rate of GDP, 2002 (%)
Saint Lucia	61	149	244.3	0.7	69.5	4 611	0.0
Saint Pierre and Miquelon	23	6	26.1	0.3	41.7	-	-
Saint Vincent and Grenadines	39	120	307.7	0.8	11.0	3 137	1.1
Trinidad and Tobago	513	1 303	254.0	0.3	24.6	7 607	2.7
United States	915 895	294 043	32.1	1.0	19.9	36 924	2.4
United States Virgin Islands	34	111	326.5	0.9	6.4	-	-
Oceania, total	**849 096**	**32 234**	**3.8**	**1.2**	**26.9**	**...**	**...**
American Samoa	20	62	310.0	3.7	9.7	-	-
Australia	768 230	19 731	2.6	1.0	8.0	26 525	2.7
Cook Islands	23	18	78.3	0.6	29.8	7 332	-
Fiji	1 827	839	45.9	1.0	48.3	2 761	4.1
French Polynesia	366	244	66.7	1.6	47.9	17 918	-
Guam	55	163	296.4	1.2	6.3	-	-
Kiribati	73	88	120.5	1.4	52.7	781	2.8
Marshall Islands	18	53	294.4	3.3	33.7	2 108	4.0
Micronesia	69	526	762.3	1.7	30.9	2 281	0.8
Nauru	2	13	650.0	1.9	0.0	3 465	-
New Caledonia	1 828	228	12.5	1.8	38.8	16 751	-
New Zealand	26 799	3 875	14.5	0.8	14.1	19 350	4.3
Niue	26	2	7.7	-1.9	64.5	-	-
Northern Mariana Islands	46	79	171.7	5.9	5.8	-	-
Palau	46	20	43.5	2.4	31.4	6 174	3.0
Papua New Guinea	45 239	5 711	12.6	2.2	86.8	577	-0.5
Samoa	282	178	63.1	1.1	77.7	1 807	1.9
Solomon Islands	2 856	477	16.7	2.9	83.5	568	-2.7
Tonga	73	104	142.5	1.0	66.6	1 626	1.6
Vanuatu	1 218	212	17.4	2.4	77.2	1 140	-0.3
South America, total	**1 754 741**	**362 277**	**20.6**	**1.4**	**18.9**	**...**	**...**
Argentina	273 669	38 428	14.0	1.2	9.9	3 375	-10.9
Bolivia	108 438	8 808	8.1	1.9	36.6	878	2.8
Brazil	845 651	178 470	21.1	1.2	16.9	2 700	1.5
Chile	74 881	15 805	21.1	1.2	13.0	4 523	2.1
Colombia	103 871	44 222	42.6	1.6	23.5	1 744	1.6
Ecuador	27 684	13 003	47.0	1.5	38.2	2 108	3.4
Falkland Islands	1 217	3	0.2	0.5	17.3	-	-
French Guiana	8 815	178	2.0	2.2	24.6	9 705	-
Guyana	21 498	765	3.6	0.3	62.4	1 010	-1.1
Paraguay	39 730	5 878	14.8	2.3	42.8	1 001	-2.3
Peru	128 000	27 167	21.2	1.5	26.1	2 238	4.9
Suriname	15 600	436	2.8	0.7	23.9	2 240	3.0
Uruguay	17 481	3 415	19.5	0.7	7.4	3 274	-10.8
Venezuela	88 206	25 699	29.1	1.8	12.3	2 994	-8.9
World, total	**13 063 900**	**6 301 463**	**48.2**	**1.2**	**51.7**	**...**	**...**

Note: The regional breakdown reflects geographic rather than economic or political groupings.

Country/area	Total land area ('000 ha)	Forest area, 2000				Forest cover change, 1990–2000	
		Total forest ('000 ha)	% of land area	Area per capita (ha)	Forest plantations ('000 ha)	Annual change ('000 ha)	Annual rate of change (%)
Estonia	4 227	2 060	48.7	1.5	305	13	0.6
Finland	30 459	21 935	72.0	4.2	0	8	n.s.
France	55 010	15 341	27.9	0.3	961	62	0.4
Germany	34 927	10 740	30.7	0.1	0	n.s.	n.s.
Greece	12 890	3 599	27.9	0.3	120	30	0.9
Hungary	9 234	1 840	19.9	0.2	136	7	0.4
Iceland	10 025	31	0.3	0.1	12	1	2.2
Ireland	6 889	659	9.6	0.2	590	17	3.0
Italy	29 406	10 003	34.0	0.2	133	30	0.3
Latvia	6 205	2 923	47.1	1.2	143	13	0.4
Liechtenstein	15	7	46.7	0.2	0	n.s.	1.2
Lithuania	6 258	1 994	31.9	0.5	284	5	0.2
Malta	32	n.s.	n.s.	-	n.s.	n.s.	n.s.
Netherlands	3 392	375	11.1	n.s.	100	1	0.3
Norway	30 683	8 868	28.9	2.0	300	31	0.4
Poland	30 442	9 047	29.7	0.2	39	18	0.2
Portugal	9 150	3 666	40.1	0.4	834	57	1.7
Republic of Moldova	3 296	325	9.9	0.1	1	1	0.2
Romania	23 034	6 448	28.0	0.3	91	15	0.2
Russian Federation	1 688 851	851 392	50.4	5.8	17 340	135	n.s.
San Marino	6	-	-	-	-	-	-
Serbia and Montenegro	10 200	2 887	28.3	0.3	39	-1	-0.1
Slovakia	4 808	2 177	45.3	0.4	15	18	0.9
Slovenia	2 012	1 107	55.0	0.6	1	2	0.2
Spain	49 945	14 370	28.8	0.4	1 904	86	0.6
Sweden	41 162	27 134	65.9	3.1	569	1	n.s.
Switzerland	3 955	1 199	30.3	0.2	4	4	0.4
The FYR of Macedonia	2 543	906	35.6	0.5	30	n.s.	n.s.
Ukraine	57 935	9 584	16.5	0.2	4 425	31	0.3
United Kingdom	24 160	2 794	11.6	n.s.	1 928	17	0.6
North and Central America, total	**2 136 966**	**549 304**	**25.7**	**1.1**	**17 533**	**-570**	**-0.1**
Antigua and Barbuda	44	9	20.5	0.1	0	n.s.	n.s.
Bahamas	1 001	842	84.1	2.8	-	n.s.	n.s.
Barbados	43	2	4.7	n.s.	0	n.s.	n.s.
Belize	2 280	1 348	59.1	5.7	3	-36	-2.3
Bermuda	5	-	-	-	-	-	-
British Virgin Islands	15	3	20.0	0.1	-	n.s.	n.s.
Canada	922 097	244 571	26.5	7.9	0	n.s.	n.s.
Cayman Islands	26	13	-	0.4	-	n.s.	n.s.
Costa Rica	5 106	1 968	38.5	0.5	178	-16	-0.8
Cuba	10 982	2 348	21.4	0.2	482	28	1.3
Dominica	75	46	61.3	0.6	n.s.	n.s.	-0.7
Dominican Republic	4 838	1 376	28.4	0.2	30	n.s.	n.s.
El Salvador	2 072	121	5.8	n.s.	14	-7	-4.6
Greenland	34 170	-	-	-	-	-	-
Grenada	34	5	14.7	0.1	n.s.	n.s.	0.9
Guadeloupe	169	82	48.5	0.2	4	2	2.1
Guatemala	10 843	2 850	26.3	0.3	133	-54	-1.7
Haiti	2 756	88	3.2	n.s.	20	-7	-5.7
Honduras	11 189	5 383	48.1	0.9	48	-59	-1.0
Jamaica	1 083	325	30.0	0.1	9	-5	-1.5
Martinique	107	47	43.9	0.1	2	n.s.	n.s.
Mexico	190 869	55 205	28.9	0.6	267	-631	-1.1
Montserrat	11	3	27.3	0.3	-	n.s.	n.s.
Netherlands Antilles	80	1	n.s.	n.s.	-	n.s.	n.s.
Nicaragua	12 140	3 278	27.0	0.7	46	-117	-3.0
Panama	7 443	2 876	38.6	1.0	40	-52	-1.6
Puerto Rico	887	229	25.8	0.1	4	-1	-0.2
Saint Kitts and Nevis	36	4	11.1	0.1	0	n.s.	-0.6

Note: The regional breakdown reflects geographic rather than economic or political groupings.

Country/area	Total land area ('000 ha)	Forest area, 2000				Forest cover change, 1990–2000	
		Total forest ('000 ha)	% of land area	Area per capita (ha)	Forest plantations ('000 ha)	Annual change ('000 ha)	Annual rate of change (%)
Saint Lucia	61	9	14.8	0.1	1	-1	-4.9
Saint Pierre and Miquelon	23	-	-	-	-	-	-
Saint Vincent and Grenadines	39	6	15.4	0.1	0	n.s.	-1.4
Trinidad and Tobago	513	259	50.5	0.2	15	-2	-0.8
United States	915 895	225 993	24.7	0.8	16 238	388	0.2
United States Virgin Islands	34	14	41.2	0.1	-	n.s.	n.s.
Oceania, total	**849 096**	**197 623**	**23.3**	**6.6**	**2 848**	**-365**	**-0.2**
American Samoa	20	12	60.1	0.2	0	n.s.	n.s.
Australia	768 230	154 539	20.1	8.3	1 043	-282	-0.2
Cook Islands	23	22	95.7	1.2	1	n.s.	n.s.
Fiji	1 827	815	44.6	1.0	97	-2	-0.2
French Polynesia	366	105	28.7	0.5	5	n.s.	n.s.
Guam	55	21	38.2	0.1	n.s.	n.s.	n.s.
Kiribati	73	28	38.4	0.3	0	n.s.	n.s.
Marshall Islands	18	n.s.	-	-	-	n.s.	n.s.
Micronesia	69	15	21.7	0.1	n.s.	-1	-4.5
Nauru	2	n.s.	-	-	-	n.s.	n.s.
New Caledonia	1 828	372	20.4	1.8	10	n.s.	n.s.
New Zealand	26 799	7 946	29.7	2.1	1 542	39	0.5
Niue	26	6	-	3.0	n.s.	n.s.	n.s.
Northern Mariana Islands	46	14	30.4	0.2	-	n.s.	n.s.
Palau	46	35	76.1	1.8	n.s.	n.s.	n.s.
Papua New Guinea	45 239	30 601	67.6	6.5	90	-113	-0.4
Samoa	282	105	37.2	0.6	5	-3	-2.1
Solomon Islands	2 856	2 536	88.8	5.9	50	-4	-0.2
Tonga	73	4	5.5	n.s.	1	n.s.	n.s.
Vanuatu	1 218	447	36.7	2.4	3	1	0.1
South America, total	**1 754 741**	**885 618**	**50.5**	**2.6**	**10 455**	**-3 711**	**-0.4**
Argentina	273 669	34 648	12.7	0.9	926	-285	-0.8
Bolivia	108 438	53 068	48.9	6.5	46	-161	-0.3
Brazil	845 651	543 905	64.3	3.2	4 982	-2 309	-0.4
Chile	74 881	15 536	20.7	1.0	2 017	-20	-0.1
Colombia	103 871	49 601	47.8	1.2	141	-190	-0.4
Ecuador	27 684	10 557	38.1	0.9	167	-137	-1.2
Falkland Islands	1 217	-	-	-	-	-	-
French Guiana	8 815	7 926	89.9	45.6	1	n.s.	n.s.
Guyana	21 498	16 879	78.5	19.7	12	-49	-0.3
Paraguay	39 730	23 372	58.8	4.4	27	-123	-0.5
Peru	128 000	65 215	50.9	2.6	640	-269	-0.4
Suriname	15 600	14 113	90.5	34.0	13	n.s.	n.s.
Uruguay	17 481	1 292	7.4	0.4	622	50	5.0
Venezuela	88 206	49 506	56.1	2.1	863	-218	-0.4
World, total	**13 063 900**	**3 869 455**	**29.6**	**0.6**	**186 733**	**-9 391**	**-0.2**

Note: The regional breakdown reflects geographic rather than economic or political groupings.

TABLE 3
Forest types, volume and biomass

Country/area	Forest types (% of country's forest area)				Wood volume in forests		Wood biomass in forests	
	Tropical	Subtropical	Temperate	Boreal/polar	(m³/ha)	Total (million m³)	(tonnes/ha)	Total (million tonnes)
Africa, total	**98**	**1**	**0**	**0**	**72**	**46 472**	**109**	**70 917**
Algeria	0	100	0	0	44	94	75	160
Angola	100	0	0	0	39	2 714	54	3 774
Benin	100	0	0	0	140	371	195	518
Botswana	100	0	0	0	45	560	63	779
Burkina Faso	100	0	0	0	10	74	16	113
Burundi	100	0	0	0	110	10	187	18
Cameroon	100	0	0	0	135	3 211	131	3 129
Cape Verde	100	0	0	0	83	7	127	11
Central African Republic	100	0	0	0	85	1 937	113	2 583
Chad	100	0	0	0	11	134	16	205
Comoros	100	0	0	0	60	0	65	1
Congo	100	0	0	0	132	2 916	213	4 699
Cote d'Ivoire	100	0	0	0	133	948	130	924
Dem. Rep. of the Congo	100	0	0	0	133	17 932	225	30 403
Djibouti	100	0	0	0	21	0	46	0
Egypt	0	100	0	0	108	8	106	8
Equatorial Guinea	100	0	0	0	93	163	158	277
Eritrea	100	0	0	0	23	36	32	50
Ethiopia	100	0	0	0	56	259	79	363
Gabon	100	0	0	0	128	2 791	137	2 991
Gambia	100	0	0	0	13	6	22	11
Ghana	100	0	0	0	49	311	88	556
Guinea	100	0	0	0	117	808	114	788
Guinea-Bissau	100	0	0	0	19	41	20	44
Kenya	100	0	0	0	35	593	48	826
Lesotho	0	100	0	0	34	0	34	0
Liberia	100	0	0	0	201	699	196	681
Libyan Arab Jamahiriya	0	100	0	0	14	5	20	7
Madagascar	100	0	0	0	114	1 339	194	2 270
Malawi	100	0	0	0	103	264	143	365
Mali	100	0	0	0	22	289	31	402
Mauritania	100	0	0	0	4	1	6	2
Mauritius	100	0	0	0	88	1	95	2
Morocco	0	100	0	0	27	80	41	123
Mozambique	100	0	0	0	25	774	55	1 683
Namibia	100	0	0	0	7	54	12	94
Niger	100	0	0	0	3	4	4	6
Nigeria	100	0	0	0	82	1 115	184	2 493
Réunion	100	0	0	0	115	8	160	11
Rwanda	100	0	0	0	110	34	187	58
Saint Helena	100	0	0	0	-	-	-	-
Sao Tome and Principe	100	0	0	0	108	3	116	3
Senegal	100	0	0	0	31	192	30	187
Seychelles	100	0	0	0	29	1	49	1
Sierra Leone	100	0	0	0	143	151	139	147
Somalia	100	0	0	0	18	138	26	192
South Africa	68	32	0	0	49	437	81	720
Sudan	100	0	0	0	9	531	12	740
Swaziland	86	14	0	0	39	20	115	60
Togo	100	0	0	0	92	47	155	79
Tunisia	0	100	0	0	18	9	27	14
Uganda	100	0	0	0	133	559	163	681
United Republic of Tanzania	100	0	0	0	43	1 676	60	2 333
Western Sahara	100	0	0	0	18	3	59	9
Zambia	100	0	0	0	43	1 347	104	3 262
Zimbabwe	100	0	0	0	40	765	56	1 065

Note: The regional breakdown reflects geographic rather than economic or political groupings.

Country/area	Forest types (% of country's forest area)				Wood volume in forests		Wood biomass in forests	
	Tropical	Subtropical	Temperate	Boreal/polar	*(m³/ha)*	Total *(million m³)*	*(tonnes/ha)*	Total *(million tonnes)*
Asia, total	**61**	**23**	**14**	**2**	**63**	**34 506**	**82**	**45 062**
Afghanistan	0	100	0	0	22	30	27	37
Armenia	0	61	39	0	128	45	66	23
Azerbaijan	0	38	62	0	136	149	105	115
Bahrain	0	100	0	0	14	-	14	-
Bangladesh	100	0	0	0	23	31	39	52
Bhutan	69	31	0	0	163	492	178	537
Brunei Darussalam	100	0	0	0	119	52	205	90
Cambodia	100	0	0	0	40	376	69	648
China	3	59	29	8	52	8 437	61	10 038
Cyprus	0	100	0	0	43	7	21	4
Dem. People's Rep. of Korea	0	0	100	0	41	333	25	209
Georgia	0	41	59	0	145	434	97	291
India	95	5	0	0	43	2 730	73	4 706
Indonesia	100	0	0	0	79	8 242	136	14 226
Iran, Islamic Rep. of	0	98	2	0	86	631	149	1 089
Iraq	0	100	0	0	29	23	28	22
Israel	0	100	0	0	49	6	-	-
Japan	0	54	46	0	145	3 485	88	2 128
Jordan	0	100	0	0	38	3	37	3
Kazakhstan	0	0	83	17	35	428	18	214
Kuwait	0	100	0	0	21	0	21	0
Kyrgyzstan	0	0	100	0	32	32	-	-
Lao People's Dem. Rep.	100	0	0	0	29	359	31	391
Lebanon	0	100	0	0	23	1	22	1
Malaysia	100	0	0	0	119	2 288	205	3 949
Maldives	100	0	0	0	-	-	-	-
Mongolia	0	0	100	0	128	1 359	80	853
Myanmar	99	1	0	0	33	1 137	57	1 965
Nepal	58	42	0	0	100	391	109	427
Oman	100	0	0	0	17	0	17	0
Pakistan	2	98	0	0	22	53	27	64
Palestine, Occupied Territory	0	0	0	0	-	-	-	-
Philippines	100	0	0	0	66	383	114	661
Qatar	0	100	0	0	13	0	12	0
Republic of Korea	0	15	85	0	58	362	36	227
Saudi Arabia	91	9	0	0	12	18	12	18
Singapore	100	0	0	0	119	0	205	0
Sri Lanka	100	0	0	0	34	66	59	114
Syrian Arab Republic	0	100	0	0	29	13	28	13
Tajikistan	0	0	100	0	14	6	10	4
Thailand	100	0	0	0	17	252	29	434
Timor-Leste	100	0	0	0	79	40	136	69
Turkey	0	92	8	0	136	1 386	74	754
Turkmenistan	0	4	96	0	4	14	3	10
United Arab Emirates	100	0	0	0	-	-	-	-
Uzbekistan	0	0	100	0	6	11	-	-
Viet Nam	98	2	0	0	38	372	66	643
Yemen	100	0	0	0	14	6	19	9
Europe, total	**0**	**5**	**22**	**73**	**112**	**116 448**	**59**	**61 070**
Albania	0	83	17	0	81	80	58	57
Andorra	0	0	100	0	0	-	0	-
Austria	0	0	100	0	286	1 110	250	970
Belarus	0	0	100	0	153	1 436	80	755
Belgium and Luxembourg	0	0	100	0	218	159	101	74
Bosnia and Herzegovina	0	19	81	0	110	250	-	-
Bulgaria	0	6	94	0	130	480	76	279
Croatia	0	28	71	0	201	358	107	190
Czech Republic	0	0	100	0	260	684	125	329
Denmark	0	0	100	0	124	56	58	26
Estonia	0	0	100	0	156	321	85	175

Note: The regional breakdown reflects geographic rather than economic or political groupings.

Country/area	Forest types (% of country's forest area)				Wood volume in forests		Wood biomass in forests	
	Tropical	Subtropical	Temperate	Boreal/polar	(m³/ha)	Total (million m³)	(tonnes/ha)	Total (million tonnes)
Finland	0	0	2	98	89	1 945	50	1 089
France	0	0	100	0	191	2 927	92	1 418
Germany	0	0	100	0	268	2 880	134	1 440
Greece	0	97	3	0	45	163	25	90
Hungary	0	0	100	0	174	320	112	207
Iceland	0	0	0	100	27	1	17	1
Ireland	0	0	100	0	74	49	25	16
Italy	0	84	16	0	145	1 450	74	742
Latvia	0	0	100	0	174	509	93	272
Liechtenstein	0	0	100	0	254	2	119	1
Lithuania	0	0	100	0	183	366	99	197
Malta	0	100	0	0	232	0	-	-
Netherlands	0	0	100	0	160	60	107	40
Norway	0	0	7	93	89	785	49	432
Poland	0	0	100	0	213	1 930	94	851
Portugal	0	81	19	0	82	299	33	120
Republic of Moldova	0	0	100	0	128	42	64	21
Romania	0	0	100	0	213	1 373	124	801
Russian Federation	0	0	14	86	105	89 136	56	47 423
San Marino	0	0	0	0	0	-	0	-
Serbia and Montenegro	0	16	84	0	111	321	23	67
Slovakia	0	0	100	0	253	552	142	308
Slovenia	0	12	88	0	283	313	178	197
Spain	0	80	20	0	44	632	24	347
Sweden	0	0	28	72	107	2 914	63	1 722
Switzerland	0	0	100	0	337	404	165	198
The FYR of Macedonia	0	64	36	0	70	63	-	-
Ukraine	0	0	100	0	179	1 719	-	-
United Kingdom	0	0	87	13	128	359	76	213
North and Central America, total	**15**	**16**	**29**	**40**	**123**	**67 329**	**95**	**52 357**
Antigua and Barbuda	100	0	0	0	116	1	210	2
Bahamas	100	0	0	0	-	-	-	-
Barbados	100	0	0	0	-	-	-	-
Belize	100	0	0	0	202	272	211	284
Bermuda	0	0	0	0	-	-	-	-
British Virgin Islands	100	0	0	0	-	-	-	-
Canada	0	0	26	74	120	29 364	83	20 240
Cayman Islands	100	0	0	0	-	-	-	-
Costa Rica	100	0	0	0	211	414	220	433
Cuba	100	0	0	0	71	167	114	268
Dominica	100	0	0	0	91	4	166	8
Dominican Republic	100	0	0	0	29	40	53	73
El Salvador	100	0	0	0	223	27	202	24
Greenland	0	0	0	0	-	-	-	-
Grenada	100	0	0	0	83	0	150	1
Guadeloupe	100	0	0	0	-	-	-	-
Guatemala	100	0	0	0	355	1 012	371	1 057
Haiti	100	0	0	0	28	2	101	9
Honduras	100	0	0	0	58	311	105	566
Jamaica	100	0	0	0	82	27	171	56
Martinique	100	0	0	0	5	0	5	0
Mexico	70	30	0	0	52	2 871	54	2 981
Montserrat	100	0	0	0	-	-	-	-
Netherlands Antilles	100	0	0	0	-	-	-	-
Nicaragua	100	0	0	0	154	506	161	528
Panama	100	0	0	0	308	887	322	926
Puerto Rico	100	0	0	0	-	-	-	-

Note: The regional breakdown reflects geographic rather than economic or political groupings.

Country/area	Forest types (% of country's forest area)				Wood volume in forests		Wood biomass in forests	
	Tropical	Subtropical	Temperate	Boreal/polar	(m³/ha)	Total (million m³)	(tonnes/ha)	Total (million tonnes)
Saint Kitts and Nevis	100	0	0	0	-	-	-	-
Saint Lucia	100	0	0	0	190	2	198	2
Saint Pierre and Miquelon	0	0	0	100	-	-	-	-
Saint Vincent and Grenadines	100	0	0	0	166	1	173	1
Trinidad and Tobago	100	0	0	0	71	18	129	33
United States	0	37	48	15	136	30 838	108	24 428
United States Virgin Islands	100	0	0	0	-	-	-	-
Oceania, total	**62**	**30**	**8**	**0**	**55**	**10 771**	**64**	**12 640**
American Samoa	100	0	0	0	-	-	-	-
Australia	54	38	8	0	55	8 506	57	8 840
Cook Islands	100	0	0	0	-	-	-	-
Fiji	100	0	0	0	-	-	-	-
French Polynesia	100	0	0	0	-	-	-	-
Guam	100	0	0	0	-	-	-	-
Kiribati	100	0	0	0	-	-	-	-
Marshall Islands	100	0	0	0	-	-	-	-
Micronesia	100	0	0	0	-	-	-	-
Nauru	100	0	0	0	-	-	-	-
New Caledonia	100	0	0	0	-	-	-	-
New Zealand	0	51	49	0	125	992	217	1 726
Niue	100	0	0	0	-	-	-	-
Northern Mariana Islands	100	0	0	0	-	-	-	-
Palau	100	0	0	0	-	-	-	-
Papua New Guinea	100	0	0	0	34	1 025	58	1 784
Samoa	100	0	0	0	-	-	-	-
Solomon Islands	100	0	0	0	-	-	-	-
Tonga	100	0	0	0	-	-	-	-
Vanuatu	100	0	0	0	-	-	-	-
South America, total	**96**	**2**	**1**	**0**	**125**	**110 826**	**203**	**180 210**
Argentina	91	5	4	0	25	866	68	2 356
Bolivia	100	0	0	0	114	6 050	183	9 711
Brazil	98	2	0	0	131	71 252	209	113 676
Chile	0	54	45	0	160	2 486	268	4 164
Colombia	100	0	0	0	108	5 359	196	9 722
Ecuador	100	0	0	0	121	1 275	151	1 594
Falkland Islands	0	0	100	0	-	-	-	-
French Guiana	100	0	0	0	145	1 151	253	2 003
Guyana	100	0	0	0	145	2 451	253	4 264
Paraguay	100	0	0	0	34	792	59	1 379
Peru	100	0	0	0	158	10 304	245	15 978
Suriname	100	0	0	0	145	2 049	253	3 566
Uruguay	0	100	0	0	-	-	-	-
Venezuela	100	0	0	0	134	6 629	233	11 535
World, total	**52**	**9**	**13**	**25**	**100**	**386 352**	**109**	**422 256**

Note: The regional breakdown reflects geographic rather than economic or political groupings.

TABLE 4
Production, trade and consumption of forest products, 2002

Country/area	Woodfuel ('000 m³)				Industrial roundwood ('000 m³)				Sawnwood ('000 m³)			
	Production	Imports	Exports	Consumption	Production	Imports	Exports	Consumption	Production	Imports	Exports	Consump
Africa, total	**545 915**	**1**	**1**	**545 915**	**66 785**	**361**	**5 481**	**61 665**	**7 761**	**4 083**	**1 920**	**9 9**
Algeria	7 305	-	-	7 305	208	23	0	231	13	935	0	9
Angola	3 320	0	-	3 320	1 116	0	1	1 115	5	0	0	
Benin	5 966	0	-	5 966	332	66	0	398	46	0	2	
Botswana	645	0	-	645	105	-	-	105	-	15	-	
Burkina Faso	11 400	-	-	11 400	594	0	1	593	1	19	0	
Burundi	8 095	-	-	8 095	333	-	4	329	83	0	0	
Cameroon	9 256	-	-	9 256	1 270	0	219	1 051	800	0	795	
Cape Verde	-	0	-	0	-	1	0	1	-	1	0	
Central African Republic	2 000	-	-	2 000	1 058	0	100	958	150	0	76	
Chad	6 119	-	-	6 119	761	-	-	761	2	17	1	
Comoros	-	0	-	0	9	0	0	9	-	1	0	
Congo	1 186	-	-	1 186	1 251	0	559	692	140	0	135	
Côte d'Ivoire	8 581	-	-	8 581	3 000	10	86	2 998	620	0	349	2
Dem. Republic of the Congo	67 285	-	0	67 285	3 653	0	2	3 651	40	0	2	
Djibouti	0	0	-	0	0	0	-	0	-	1	0	
Egypt	16 484	0	0	16 484	268	116	0	384	4	1 564	0	1 5
Equatorial Guinea	447	-	-	447	364	0	515	0	4	0	2	
Eritrea	2 323	-	-	2 323	2	-	-	2	-	1	0	
Ethiopia	90 202	-	-	90 202	2 458	0	0	2 458	60	8	0	
Gabon	520	-	-	520	2 584	0	2 500	84	117	0	103	
Gambia	620	-	-	620	113	0	0	113	1	1	0	
Ghana	20 678	-	0	20 678	1 104	0	0	1 104	461	0	207	2
Guinea	11 537	-	-	11 537	651	0	32	619	26	1	0	
Guinea-Bissau	422	-	0	422	170	0	7	163	16	0	0	
Kenya	20 002	0	0	20 002	1 977	0	0	1 977	185	1	1	1
Lesotho	2 034	0	-	2 034	-	-	-	0	-	0	-	
Liberia	5 133	-	-	5 133	337	0	1 000	0	30	0	26	
Libyan Arab Jamahiriya	536	0	-	536	116	1	0	117	31	82	0	1
Madagascar	10 202	-	-	10 202	97	0	14	83	95	0	26	
Malawi	5 029	-	0	5 029	520	0	0	520	45	0	3	
Mali	4 846	0	-	4 846	413	3	3	413	13	0	0	
Mauritania	1 502	-	-	1 502	6	0	0	6	-	0	0	
Mauritius	9	0	0	9	8	7	0	15	3	43	0	
Morocco	400	0	0	400	526	88	0	614	83	452	0	5
Mozambique	16 724	0	0	16 724	1 319	0	13	1 306	28	0	5	
Namibia	-	-	-	0	-	5	-	5	-	9	-	
Niger	8 190	-	-	8 190	411	0	0	411	4	1	5	
Nigeria	60 064	-	1	60 063	9 418	1	4	9 415	2 000	1	0	2 0
Réunion	31	-	-	31	5	1	2	4	2	85	38	
Rwanda	7 500	-	-	7 500	336	0	0	336	79	0	0	
Saint Helena	-	-	-	0	-	0	-	0	-	0	0	
Sao Tome and Principe	-	0	-	0	9	0	0	9	5	0	0	
Senegal	5 178	-	-	5 178	794	23	0	817	23	86	1	1
Seychelles	-	-	-	0	-	0	0	0	-	0	1	
Sierra Leone	5 374	-	-	5 374	124	0	0	124	5	2	0	
Somalia	9 827	-	0	9 827	110	1	4	107	14	0	0	
South Africa	12 000	0	0	12 000	18 616	1	386	18 231	1 498	380	0	1 8
Sudan	17 068	0	0	17 068	2 173	0	0	2 173	51	53	110	
Swaziland	560	-	-	560	330	0	0	330	102	0	2	1
Togo	5 600	-	-	5 600	208	1	17	192	13	10	6	
Tunisia	2 116	0	-	2 116	214	15	0	229	20	311	0	3
Uganda	35 142	-	-	35 142	3 175	0	0	3 175	264	1	0	2
United Republic of Tanzania	21 125	0	0	21 125	2 314	0	10	2 304	24	0	1	
Western Sahara	-	-	-	0	-	-	-	0	-	0	-	
Zambia	7 219	0	-	7 219	834	1	1	834	157	1	1	1
Zimbabwe	8 115	0	0	8 115	992	0	0	992	397	0	25	3

Note: The regional breakdown reflects geographic rather than economic or political groupings.

Wood-based panels ('000 m³)				Pulp for paper ('000 tonnes)				Paper and paperboard ('000 tonnes)				Country/area
Production	Imports	Exports	Consumption	Production	Imports	Exports	Consumption	Production	Imports	Exports	Consumption	
2 254	**727**	**809**	**2 172**	**2 550**	**277**	**501**	**2 326**	**3 277**	**1 693**	**689**	**4 281**	**Africa, total**
50	133	0	183	2	17	0	19	41	180	0	221	Algeria
11	1	0	12	15	0	0	15	0	2	0	2	Angola
-	1	0	1	-	0	0	0	-	4	0	4	Benin
-	0	-	0	-	-	-	0	-	10	-	10	Botswana
-	2	0	2	-	0	-	0	-	8	0	8	Burkina Faso
-	1	0	1	-	0	-	0	-	0	0	0	Burundi
98	0	74	24	0	1	-	1	0	38	0	38	Cameroon
-	10	0	10	-	-	-	0	-	1	0	1	Cape Verde
4	0	0	4	-	-	0	0	-	0	0	0	Central African Republic
-	0	0	0	-	0	0	0	-	1	0	1	Chad
-	0	0	0	-	0	-	0	-	0	-	0	Comoros
25	0	22	3	-	0	0	0	-	2	0	2	Congo
323	0	189	134	-	0	-	0	-	69	0	69	Côte d'Ivoire
3	0	0	3	-	0	-	0	3	2	0	5	Dem. Republic of the Congo
-	6	0	6	-	3	0	3	-	3	0	3	Djibouti
131	161	3	289	120	90	-	210	460	522	48	934	Egypt
15	0	11	4	-	0	-	0	-	0	0	0	Equatorial Guinea
-	15	-	15	-	-	-	0	-	1	-	1	Eritrea
22	24	-	46	9	12	0	21	11	20	0	31	Ethiopia
251	0	165	86	-	0	-	0	-	6	0	6	Gabon
-	0	0	0	-	-	-	0	-	0	0	0	Gambia
391	0	192	199	-	0	0	0	-	45	0	45	Ghana
0	5	0	5	-	0	-	0	-	4	0	4	Guinea
-	0	0	0	-	-	-	0	-	0	0	0	Guinea-Bissau
52	2	3	51	66	2	0	68	129	15	7	137	Kenya
-	-	-	0	-	-	-	0	-	-	-	0	Lesotho
30	0	0	30	-	0	-	0	-	0	-	0	Liberia
-	23	0	23	-	2	0	2	6	11	0	17	Libyan Arab Jamahiriya
5	1	0	6	3	0	0	3	3	6	0	9	Madagascar
18	3	0	21	-	0	-	0	-	4	0	4	Malawi
-	1	6	0	-	0	-	0	-	3	0	3	Mali
-	0	1	0	-	0	-	0	-	2	0	2	Mauritania
0	51	0	51	-	1	-	1	-	39	2	37	Mauritius
35	5	10	30	112	18	78	52	129	125	3	251	Morocco
3	4	17	0	-	0	0	0	0	1	0	1	Mozambique
-	-	-	0	-	-	-	0	-	15	-	15	Namibia
-	0	0	0	-	7	0	7	-	1	0	1	Niger
95	74	0	169	23	17	-	40	19	182	2	199	Nigeria
-	24	0	24	-	0	0	0	-	15	0	15	Réunion
0	4	-	4	-	-	0	0	-	1	0	1	Rwanda
-	0	-	0	-	-	-	0	-	0	-	0	Saint Helena
-	0	0	0	-	0	-	0	-	0	0	0	Sao Tome and Principe
-	11	0	11	-	0	-	0	-	31	2	29	Senegal
-	1	-	1	-	-	-	0	-	0	0	0	Seychelles
-	8	0	8	-	0	1	0	-	1	1	1	Sierra Leone
0	1	0	1	-	0	0	0	-	2	0	2	Somalia
476	99	101	474	1 903	63	223	1 743	2 267	166	613	1 820	South Africa
2	5	0	7	-	0	-	0	3	6	0	9	Sudan
8	-	0	8	191	-	191	0	-	-	-	0	Swaziland
-	1	0	1	-	0	-	0	-	3	0	3	Togo
104	37	1	140	14	34	8	40	94	97	4	187	Tunisia
5	0	0	5	-	0	-	0	3	3	0	6	Uganda
4	1	1	4	54	0	0	54	25	8	1	32	United Republic of Tanzania
-	-	-	0	-	-	-	0	-	-	-	0	Western Sahara
18	7	0	25	-	0	-	0	4	4	0	8	Zambia
77	3	10	70	42	9	0	51	80	35	5	110	Zimbabwe

Country/area	Woodfuel ('000 m³)				Industrial roundwood ('000 m³)				Sawnwood ('000 m³)			
	Production	Imports	Exports	Consumption	Production	Imports	Exports	Consumption	Production	Imports	Exports	Consump...
Asia, total	**782 160**	**258**	**23**	**782 395**	**222 563**	**51 346**	**8 034**	**265 875**	**61 157**	**24 205**	**8 136**	**77 2**
Afghanistan	1 351	0	0	1 351	1760	0	0	1760	400	31	0	4
Armenia	46	0	0	46	8	2	1	9	4	27	16	
Azerbaijan	6	0	-	6	7	1	0	8	0	203	1	2
Bahrain	-	0	-	0	-	0	0	0	-	14	0	
Bangladesh	27 763	-	0	27 763	575	3	0	578	70	1	0	
Bhutan	4 348	-	-	4 348	134	0	0	134	31	-	0	
Brunei Darussalam	12	0	0	12	217	0	1	216	90	0	0	
Cambodia	9 737	-	0	9 737	125	-	0	125	5	0	5	
China	191 047	7	6	191 048	93 121	25 857	695	118 283	9 431	6 914	657	15 6
Cyprus	5	0	0	5	10	3	0	13	7	77	0	
Dem. People's Rep. of Korea	5 620	-	0	5 620	1 500	0	40	1 460	280	1	22	2
Georgia	0	0	1	0	0	0	56	0	-	3	18	
India	300 564	0	0	300 564	19 308	1 998	8	21 298	7 900	30	8	7 9
Indonesia	82 556	0	1	82 555	32 997	180	502	32 675	6 500	144	2 016	4 6
Iran, Islamic Rep. of	257	0	0	257	1 060	8	0	1 068	106	127	0	2
Iraq	53	0	-	53	59	0	-	59	12	16	0	
Israel	2	0	0	2	25	86	0	111	0	409	0	4
Japan	124	1	0	125	15 092	12 662	4	27 750	14 402	8 584	22	22 9
Jordan	237	0	-	237	4	0	0	4	-	137	0	1
Kazakhstan	0	1	13	0	0	75	546	0	224	482	357	3
Kuwait	-	0	0	0	-	3	0	3	-	73	0	
Kyrgyzstan	16	0	0	16	10	2	0	12	6	37	2	
Lao People's Dem. Rep.	5 899	0	0	5 899	392	0	63	329	182	0	131	
Lebanon	82	0	0	82	7	0	0	7	9	274	26	2
Malaysia	3 228	0	0	3 228	17 913	414	5 176	13 151	4 594	657	2 550	2 7
Maldives	-	0	-	0	-	0	-	0	-	0	0	
Mongolia	186	0	0	186	445	7	1	451	300	2	3	2
Myanmar	35 403	-	0	35 403	5 539	0	877	4 662	381	0	269	1
Nepal	12 728	-	-	12 728	1 260	0	0	1 260	630	0	0	6
Oman	-	2	1	1	-	7	0	7	-	112	0	1
Pakistan	25 013	0	-	25 013	1 605	295	0	1 310	1 180	28	0	1 2
Philippines	13 328	0	0	13 328	3 079	434	1	3 512	154	401	91	4
Qatar	-	0	-	0	-	9	-	9	-	12	0	
Republic of Korea	2 458	0	0	2 458	1 592	7 657	0	9 249	5 194	848	14	6 0
Saudi Arabia	-	4	0	4	-	21	0	21	-	1 184	0	1 1
Singapore	-	1	0	1	-	34	39	0	25	224	195	
Sri Lanka	5 774	0	0	5 774	694	0	0	634	61	16	0	
Syrian Arab Republic	16	0	0	16	35	1	1	35	9	217	0	2
Tajikistan	0	0	0	0	0	0	0	0	-	40	0	
Thailand	20 250	0	0	20 250	7 800	688	0	8 488	288	1 924	1 558	6
Turkey	7 160	242	0	7 402	11 305	808	9	12 104	5 732	196	158	5 7
Turkmenistan	0	0	0	0	0	0	0	0	-	24	0	
United Arab Emirates	-	0	0	0	-	37	1	36	-	406	1	4
Uzbekistan	19	0	0	19	6	0	0	6	-	11	1	
Viet Nam	26 547	-	0	26 547	4 183	54	15	4 222	2 950	207	15	3 1
Yemen	326	0	-	326	-	0	0	0	-	110	0	
Europe, total	**106 909**	**2 037**	**3 276**	**112 222**	**480 118**	**57 053**	**71 936**	**465 235**	**127 844**	**45 119**	**58 928**	**114 0**
Albania	222	0	67	155	83	0	29	54	97	128	87	1
Andorra	-	2	0	2	-	0	0	0	-	10	0	
Austria	3 036	163	29	3 170	11 809	7 289	864	18 234	10 415	1 351	6 422	5 3
Belarus	978	1	10	969	5 969	130	1 328	4 771	2 182	152	914	1 4
Belgium	550	28	18	560	3 950	2 645	883	5 712	1 175	2 007	982	2 2
Bosnia and Herzegovina	1 150	0	0	1 150	3 076	0	0	3 076	310	21	226	1
Bulgaria	2 187	0	29	2 158	2 646	71	195	2 522	332	7	273	
Croatia	755	5	96	664	2 886	80	448	2 518	640	475	459	6
Czech Republic	1 007	3	212	798	13 534	991	2 302	12 223	3 800	381	1 448	2 7
Denmark	657	136	1	792	789	507	567	729	244	2 689	368	2 5
Estonia	1 930	0	227	1 703	8 570	639	3 132	6 077	1 900	236	1 248	8
Faeroe Islands	-	0	-	0	-	1	0	1	-	4	0	
Finland	4 482	102	4	4 580	48 529	12 586	404	60 711	13 390	257	8 187	5 4
France	2 400	26	368	2 058	33 500	1 934	3 916	31 518	10 540	3 287	1 406	12 4

Note: The regional breakdown reflects geographic rather than economic or political groupings.

Wood-based panels ('000 m³)				Pulp for paper ('000 tonnes)				Paper and paperboard ('000 tonnes)				Country/area
Production	Imports	Exports	Consumption	Production	Imports	Exports	Consumption	Production	Imports	Exports	Consumption	
8 768	19 184	16 147	61 805	40 325	12 911	2 720	50 516	97 823	20 505	11 225	107 103	**Asia, total**
1	3	0	4	-	0	0	0	-	0	0	0	Afghanistan
3	10	0	13	-	2	-	2	2	9	0	11	Armenia
1	59	1	59	0	0	0	0	143	13	0	156	Azerbaijan
-	1	0	1	-	0	-	0	-	6	0	6	Bahrain
9	6	0	15	37	5	-	42	46	38	0	84	Bangladesh
32	0	0	32	-	1	0	1	-	0	0	0	Bhutan
-	1	0	1	-	0	-	0	-	2	0	2	Brunei Darussalam
37	10	37	10	0	0	-	0	0	30	-	30	Cambodia
4 687	5 657	2 735	27 609	18 381	5 795	46	24 130	37 929	10 393	3 990	44 332	China
3	109	0	112	0	3	0	3	0	56	0	56	Cyprus
-	9	0	9	106	45	0	151	80	25	2	103	Dem. People's Rep. of Korea
10	5	0	20	-	0	0	0	-	6	0	6	Georgia
645	67	12	700	2 603	198	26	2 775	3 973	620	101	4 492	India
2 635	129	5 997	6 767	5 587	599	2 244	3 942	6 995	262	2 348	4 909	Indonesia
414	84	0	498	45	48	0	93	46	500	0	546	Iran, Islamic Rep. of
5	21	0	26	11	0	0	11	20	20	0	40	Iraq
181	210	13	378	15	125	17	123	275	455	70	660	Israel
4 893	6 342	44	11 191	10 663	2 428	107	12 984	30 686	1 805	665	31 826	Japan
-	89	0	89	8	25	0	33	25	133	27	131	Jordan
33	147	1	179	-	2	0	2	23	54	0	77	Kazakhstan
-	94	0	94	-	15	-	15	-	89	32	57	Kuwait
-	76	0	76	-	2	0	2	15	14	1	28	Kyrgyzstan
13	1	5	9	-	0	-	0	-	3	0	3	Lao People's Dem. Rep.
46	175	1	220	-	14	0	14	42	141	21	162	Lebanon
6 803	293	5 639	1 457	124	60	0	184	851	1 107	149	1 809	Malaysia
-	4	0	4	-	-	-	0	-	1	0	1	Maldives
2	4	1	5	-	0	0	0	-	5	0	5	Mongolia
20	4	79	0	47	0	0	47	42	52	0	94	Myanmar
5	4	0	9	15	0	0	15	13	2	0	15	Nepal
-	73	0	73	-	2	0	2	-	57	2	55	Oman
354	57	0	411	335	31	-	366	1 165	167	0	1 332	Pakistan
620	227	25	822	202	44	0	246	1 056	571	171	1 456	Philippines
-	6	0	6	-	0	0	0	-	3	0	3	Qatar
3 513	3 349	101	6 761	587	2 521	0	3 108	9 812	784	2 430	8 166	Republic of Korea
-	496	0	496	-	94	0	94	-	487	17	470	Saudi Arabia
355	314	147	522	-	58	86	0	87	699	163	623	Singapore
22	3	3	22	21	9	0	30	25	50	5	70	Sri Lanka
27	152	1	178	-	11	0	11	1	76	6	71	Syrian Arab Republic
-	9	-	9	-	-	-	0	-	1	0	1	Tajikistan
705	53	1 019	0	999	356	191	1 164	2 444	259	787	1 916	Thailand
2 656	333	230	2 759	278	368	2	644	1 643	1 020	175	2 488	Turkey
-	3	1	2	-	0	-	0	-	1	0	1	Turkmenistan
-	325	44	281	-	17	0	17	-	275	59	216	United Arab Emirates
-	69	0	69	-	1	0	1	-	18	0	18	Uzbekistan
40	42	9	73	314	33	0	347	384	140	2	522	Viet Nam
-	59	0	59	-	0	-	0	-	55	0	55	Yemen
3 275	24 634	27 821	60 088	47 008	17 191	12 289	51 910	102 039	47 163	58 401	90 801	**Europe, total**
37	136	1	172	0	4	0	4	3	18	1	20	Albania
-	2	0	2	-	0	-	0	-	2	0	2	Andorra
3 420	565	2 603	1 382	1 559	576	321	1 814	4 419	1 156	3 661	1 914	Austria
632	194	360	466	59	23	0	82	216	137	49	304	Belarus
2 758	1 451	2 879	1 330	490	1 008	658	840	1 710	3 080	2 244	2 546	Belgium
34	31	15	50	-	2	0	2	-	14	3	11	Bosnia and Herzegovina
533	137	280	390	102	13	60	55	171	160	52	279	Bulgaria
81	183	44	220	122	2	43	81	467	166	139	494	Croatia
1 109	611	727	993	702	153	332	523	870	670	572	968	Czech Republic
353	1 246	129	1 470	0	66	4	62	393	1 144	238	1 299	Denmark
480	134	384	230	60	0	0	60	81	87	68	100	Estonia
-	1	0	1	-	-	0	0	-	2	0	2	Faeroe Islands
1 860	261	1 500	621	11 729	96	2 114	9 711	12 776	383	11 487	1 672	Finland
5 593	1 608	270	6 931	2 561	2 224	487	4 298	9 798	5 885	4 754	10 929	France

Country/area	Woodfuel ('000 m³)				Industrial roundwood ('000 m³)				Sawnwood ('000 m³)			
	Production	Imports	Exports	Consumption	Production	Imports	Exports	Consumption	Production	Imports	Exports	Consump
Trinidad and Tobago	36	0	-	36	51	7	0	58	43	53	0	
Turks and Caicos Islands	-	0	0	0	-	2	-	2	-	4	0	
United States	73 086	0	0	73 086	404 735	6 618	11 001	400 352	89 151	37 416	4 520	122 0
United States Virgin Islands	-	-	-	0	-	0	-	0	-	-	-	
Oceania, total	**12 973**	**0**	**0**	**12 973**	**49 644**	**12**	**11 446**	**38 210**	**8 691**	**849**	**2 114**	**7 4**
American Samoa	-	-	-	0	-	0	-	0	-	1	-	
Australia	7 104	0	0	7 104	24 322	2	1 325	22 999	4 119	736	233	4 6
Cook Islands	-	-	-	0	5	0	1	4	-	3	-	
Fiji Islands	37	-	-	37	346	0	0	346	84	0	9	
French Polynesia	-	-	-	0	-	3	0	3	-	47	0	
Kiribati	-	0	-	0	-	0	-	0	-	2	-	
Marshall Islands	-	-	-	0	-	-	-	0	-	6	-	
Micronesia	-	0	-	0	-	-	-	0	-	7	-	
Nauru	-	-	-	0	-	0	-	0	-	0	0	
New Caledonia	-	0	-	0	5	1	0	4	3	3	0	
New Zealand	0	0	-	0	22 613	5	7 859	14 759	4 352	36	1 834	2 5
Niue	-	-	-	0	0	0	0	0	-	0	0	
Norfolk Island	-	-	-	0	-	0	-	0	-	1	0	
Northern Mariana Islands	-	-	-	0	-	-	-	0	-	0	-	
Palau	-	-	-	0	-	1	-	1	-	3	-	
Papua New Guinea	5 533	0	-	5 533	1 708	0	1 858	0	70	0	28	
Samoa	70	0	-	70	61	0	6	55	21	1	1	
Solomon	138	-	0	138	554	-	396	158	12	0	2	
Tonga	-	0	-	0	2	0	0	2	2	0	0	
Tuvalu	-	-	-	0	-	0	-	0	-	1	-	
Vanuatu	91	-	0	91	28	0	0	28	28	2	5	
Wallis and Futuna Islands	-	-	-	0	-	0	-	0	-	1	-	
South America, total	**189 679**	**0**	**0**	**189 679**	**153 469**	**43**	**2 670**	**150 842**	**33 183**	**299**	**5 142**	**28 3**
Argentina	3 972	0	0	3 972	5 335	0	36	5 299	2 130	24	285	1 8
Bolivia	2 184	0	0	2 184	8 054	1	2	8 053	299	1	34	2
Brazil	134 473	0	0	134 473	102 994	18	885	102 127	21 200	139	2 009	19 3
Chile	12 326	-	0	12 326	25 491	0	512	24 979	6 439	36	2 311	4 1
Colombia	9 598	0	-	9 598	2 012	0	19	1 993	527	18	16	5
Ecuador	5 274	0	-	5 274	913	0	34	879	750	0	30	7
French Guiana	84	-	-	84	60	1	2	59	15	1	4	
Guyana	873	-	0	873	269	0	48	221	35	0	33	
Paraguay	5 743	-	0	5 743	4 044	0	0	4 044	550	4	163	3
Peru	7 335	0	0	7 335	1 084	21	0	1 105	603	14	110	5
Suriname	44	0	-	44	154	0	26	128	47	0	8	
Uruguay	4 076	0	-	4 076	1 832	1	1 102	731	224	20	77	
Venezuela	3 697	0	-	3 697	1 227	0	4	1 223	364	43	61	3
World, total	**1 796 677**	**2 524**	**3 705**	**1 795 496**	**1 587 715**	**122 996**	**115 523**	**1 595 188**	**390 918**	**115 924**	**118 481**	**388 3**

Note: The regional breakdown reflects geographic rather than economic or political groupings.

Wood-based panels ('000 m³)				Pulp for paper ('000 tonnes)				Paper and paperboard ('000 tonnes)				Country/area
Production	Imports	Exports	Consumption	Production	Imports	Exports	Consumption	Production	Imports	Exports	Consumption	
-	41	0	41	0	3	0	3	-	74	1	73	Trinidad and Tobago
-	1	0	1	-	-	-	0	-	0	-	0	Turks and Caicos Islands
40 516	17 635	2 061	56 090	52 914	6 579	5 477	54 016	81 792	15 941	8 225	89 508	United States
-	-	-	0	-	-	-	0	-	-	-	0	United States Virgin Islands
3 940	**416**	**1 872**	**2 484**	**4 383**	**324**	**763**	**3 944**	**3 522**	**1 768**	**1 128**	**4 162**	**Oceania, total**
-	0	-	0	-	-	-	0	-	0	0	0	American Samoa
1 903	371	878	1 396	2 832	310	2	3 140	2 645	1 289	622	3 312	Australia
-	2	0	2	-	-	-	0	-	0	0	0	Cook Islands
10	7	6	11	-	0	-	0	-	97	1	96	Fiji Islands
-	6	0	6	-	0	-	0	-	2	0	2	French Polynesia
-	0	-	0	-	-	-	0	-	0	-	0	Kiribati
-	3	-	3	-	-	-	0	-	0	-	0	Marshall Islands
-	1	-	1	-	-	-	0	-	0	-	0	Micronesia
-	0	0	0	-	-	-	0	-	0	0	0	Nauru
-	2	0	2	-	2	-	2	-	3	0	3	New Caledonia
1 948	19	949	1 018	1 551	11	761	801	877	358	505	730	New Zealand
-	0	-	0	-	0	0	0	-	0	0	0	Niue
-	0	-	0	-	-	-	0	-	0	0	0	Norfolk Island
-	0	-	0	-	-	-	0	-	0	-	0	Northern Mariana Islands
-	1	-	1	-	-	-	0	-	0	0	0	Palau
79	1	39	40	-	-	-	0	-	16	0	16	Papua New Guinea
0	1	-	1	-	0	-	0	-	0	0	0	Samoa
0	0	-	0	-	-	-	0	-	0	0	0	Solomon
-	0	-	0	-	-	-	0	-	1	0	1	Tonga
-	0	-	0	-	-	-	0	-	0	0	0	Tuvalu
-	1	0	1	-	1	-	1	-	1	0	1	Vanuatu
-	0	-	0	-	-	-	0	-	0	-	0	Wallis and Futuna Islands
9 588	**396**	**3 054**	**6 930**	**11 539**	**830**	**4 990**	**7 379**	**11 524**	**2 114**	**1 244**	**12 394**	**South America, total**
692	10	364	338	785	74	258	601	1 338	277	116	1 499	Argentina
14	10	1	23	0	0	0	0	0	45	0	45	Bolivia
6 283	153	1 760	4 676	7 390	422	2 579	5 233	7 354	509	452	7 411	Brazil
1 543	0	607	936	2 687	29	2 152	564	1 177	318	402	1 093	Chile
182	50	40	192	358	113	1	470	837	321	119	1 039	Colombia
270	24	93	201	2	18	0	20	91	112	8	195	Ecuador
0	3	0	3	-	0	-	0	-	0	0	0	French Guiana
51	0	47	4	-	0	-	0	-	7	0	7	Guyana
161	2	70	93	-	0	0	0	13	47	3	57	Paraguay
102	61	20	143	17	35	0	52	63	249	28	284	Peru
2	3	1	4	-	0	-	0	-	2	0	2	Suriname
6	20	0	26	35	15	1	49	89	61	78	72	Uruguay
282	60	52	290	265	123	0	388	562	166	38	690	Venezuela
195 359	**65 937**	**63 953**	**197 343**	**184 715**	**39 067**	**38 418**	**185 364**	**324 649**	**95 000**	**95 425**	**324 224**	**World, total**

TABLE 5
Status of ratification of international conventions and agreements as of 1 December 2004

Country/territory	CBD	UNFCCC	Kyoto Protocol	CCD	CITES	Ramsar Convention	World Heritage Convention
Africa							
Algeria	X	X		X	X	X	X
Angola	X	X		X			X
Benin	X	X	X	X	X	X	X
Botswana	X	X	X	X	X	X	X
Burkina Faso	X	X		X	X	X	X
Burundi	X	X	X	X	X	X	X
Cameroon	X	X	X	X	X		X
Cape Verde	X	X		X			X
Central African Republic	X	X		X	X		X
Chad	X	X		X	X	X	X
Comoros	X	X		X	X	X	X
Congo	X	X		X	X	X	X
Cote d'Ivoire	X	X		X	X	X	X
Dem. Republic of the Congo	X	X		X	X	X	X
Djibouti	X	X	X	X	X	X	
Egypt	X	X		X	X	X	X
Equatorial Guinea	X	X	X	X	X	X	
Eritrea	X	X		X	X		X
Ethiopia	X	X		X	X		X
Gabon	X	X		X	X	X	X
Gambia	X	X	X	X	X	X	X
Ghana	X	X	X	X	X	X	X
Guinea	X	X	X	X	X	X	X
Guinea-Bissau	X	X		X	X	X	
Kenya	X	X		X	X	X	X
Lesotho	X	X	X	X	X	X	X
Liberia	X	X	X	X	X	X	X
Libyan Arab Jamahiriya	X	X		X	X	X	X
Madagascar	X	X	X	X	X	X	X
Malawi	X	X	X	X	X	X	X
Mali	X	X	X	X	X	X	X
Mauritania	X	X		X	X	X	X
Mauritius	X	X	X	X	X	X	X
Morocco	X	X	X	X	X	X	X
Mozambique	X	X		X	X	X	X
Namibia	X	X	X	X	X	X	X
Niger	X	X	X	X	X	X	X
Nigeria	X	X		X	X	X	X
Rwanda	X	X	X	X	X		X
Sao Tome and Principe	X	X		X	X		
Senegal	X	X	X	X	X	X	X
Seychelles	X	X	X	X	X		X
Sierra Leone	X	X		X	X	X	
Somalia				X	X		
South Africa	X	X	X	X	X	X	X
Sudan	X	X	X	X	X		X
Swaziland	X	X		X	X		
Togo	X	X	X	X	X	X	X
Tunisia	X	X	X	X	X	X	X
Uganda	X	X	X	X	X	X	X
United Republic of Tanzania	X	X	X	X	X	X	X
Zambia	X	X		X	X	X	X
Zimbabwe	X	X		X	X		X

Note: The regional breakdown reflects geographic rather than economic or political groupings.

Country/territory	CBD	UNFCCC	Kyoto Protocol	CCD	CITES	Ramsar Convention	World Heritage Convention
Asia							
Afghanistan	X	X		X	X		X
Armenia	X	X	X	X		X	X
Azerbaijan	X	X	X	X	X	X	X
Bahrain	X	X		X		X	X
Bangladesh	X	X	X	X	X	X	X
Bhutan	X	X	X	X	X		X
Brunei Darussalam				X	X		
Cambodia	X	X	X	X	X	X	X
China	X	X	X	X	X	X	X
Cyprus	X	X	X	X	X	X	X
Dem. People's Rep. of Korea	X	X		X			X
Georgia	X	X	X	X	X	X	X
India	X	X	X	X	X	X	X
Indonesia	X	X		X	X	X	X
Iran, Islamic Rep. of	X	X		X	X	X	X
Iraq							X
Israel	X	X	X	X	X	X	X
Japan	X	X	X	X	X	X	X
Jordan	X	X	X	X	X	X	X
Kazakhstan	X	X		X	X		X
Kuwait	X	X		X	X		X
Kyrgyzstan	X	X	X	X		X	X
Lao People's Dem. Rep.	X	X	X	X	X		X
Lebanon	X	X		X		X	X
Malaysia	X	X	X	X	X	X	X
Maldives	X	X	X	X			X
Mongolia	X	X	X	X	X	X	X
Myanmar	X	X	X	X	X		X
Nepal	X	X		X	X	X	X
Oman	X	X		X			X
Pakistan	X	X		X	X	X	X
Philippines	X	X	X	X	X	X	X
Qatar	X	X		X	X		X
Republic of Korea	X	X	X	X	X	X	X
Saudi Arabia	X	X		X	X		X
Singapore	X	X		X	X		
Sri Lanka	X	X	X	X	X	X	X
Syrian Arab Republic	X	X		X	X	X	X
Tajikistan	X	X		X		X	X
Thailand	X	X	X	X	X	X	X
Timor-Leste				X			
Turkey	X	X		X	X	X	X
Turkmenistan	X	X	X	X			X
United Arab Emirates	X	X		X	X		X
Uzbekistan	X	X	X	X	X	X	X
Viet Nam	X	X	X	X	X	X	X
Yemen	X	X	X	X	X		X
Europe							
Albania	X	X		X	X	X	X
Andorra				X			X
Austria	X	X	X	X	X	X	X
Belarus	X	X		X	X	X	X
Belgium	X	X	X	X	X	X	X
Bosnia and Herzegovina	X	X		X		X	X
Bulgaria	X	X	X	X	X	X	X
Croatia	X	X		X	X	X	X

Note: The regional breakdown reflects geographic rather than economic or political groupings.

Country/territory	CBD	UNFCCC	Kyoto Protocol	CCD	CITES	Ramsar Convention	World Heritage Convention
Czech Republic	X	X	X	X	X	X	X
Denmark	X	X	X	X	X	X	X
Estonia	X	X	X		X	X	X
Finland	X	X	X	X	X	X	X
France	X	X	X	X	X	X	X
Germany	X	X	X	X	X	X	X
Greece	X	X	X	X	X	X	X
Hungary	X	X	X	X	X	X	X
Iceland	X	X	X	X	X	X	X
Ireland	X	X	X	X	X	X	X
Italy	X	X	X	X	X	X	X
Latvia	X	X	X	X	X	X	X
Liechtenstein	X	X		X	X	X	
Lithuania	X	X	X	X	X	X	X
Luxembourg	X	X	X	X	X	X	X
Malta	X	X	X	X	X	X	X
Monaco	X	X		X	X	X	X
Netherlands	X	X		X	X	X	X
Norway	X	X	X	X	X	X	X
Poland	X	X	X	X	X	X	X
Portugal	X	X	X	X	X	X	X
Republic of Moldova	X	X	X	X	X	X	X
Romania	X	X	X	X	X	X	X
Russian Federation	X	X	X	X	X	X	X
San Marino	X	X		X			X
Serbia and Montenegro	X	X			X	X	X
Slovakia	X	X	X	X	X	X	X
Slovenia	X	X	X	X	X	X	X
Spain	X	X	X	X	X	X	X
Sweden	X	X	X	X	X	X	X
Switzerland	X	X	X	X	X	X	X
The FYR of Macedonia	X	X	X	X	X	X	X
Ukraine	X	X	X	X	X	X	X
United Kingdom	X	X	X	X	X	X	X
North and Central America							
Antigua and Barbuda	X	X	X	X	X		X
Bahamas	X	X	X	X	X	X	
Barbados	X	X	X	X	X		X
Belize	X	X	X	X	X	X	X
Canada	X	X	X	X	X	X	X
Cayman Islands							
Costa Rica	X	X	X	X	X	X	X
Cuba	X	X	X	X	X	X	X
Dominica	X	X		X	X		X
Dominican Republic	X	X	X	X	X	X	X
El Salvador	X	X	X	X	X	X	X
Greenland							
Grenada	X	X	X	X	X		X
Guatemala	X	X	X	X	X	X	X
Haiti	X	X		X			X
Honduras	X	X	X	X	X	X	X
Jamaica	X	X	X	X	X	X	X
Mexico	X	X	X	X	X	X	X
Nicaragua	X	X	X	X	X	X	X
Panama	X	X	X	X	X	X	X
Saint Kitts and Nevis	X	X		X	X		X
Saint Lucia	X	X	X	X	X	X	X
Saint Vincent and Grenadines	X	X		X	X		X

Note: The regional breakdown reflects geographic rather than economic or political groupings.

Country/territory	CBD	UNFCCC	Kyoto Protocol	CCD	CITES	Ramsar Convention	World Heritage Convention
Trinidad and Tobago	X	X	X	X	X	X	
United States		X		X	X	X	X
United States Virgin Islands							

Oceania

Country/territory	CBD	UNFCCC	Kyoto Protocol	CCD	CITES	Ramsar Convention	World Heritage Convention
American Samoa							
Australia	X	X		X	X	X	X
Cook Islands	X	X	X	X			
Fiji	X	X	X	X	X		X
French Polynesia							
Guam							
Kiribati	X	X	X	X			X
Marshall Islands	X	X	X	X		X	X
Micronesia	X	X	X	X			X
Nauru	X	X	X	X			
New Caledonia							
New Zealand	X	X	X	X	X	X	X
Niue	X	X	X	X			X
Northern Mariana Islands							
Palau	X	X	X	X	X	X	X
Papua New Guinea	X	X	X	X	X	X	X
Samoa	X	X	X	X	X	X	X
Solomon Islands	X	X	X	X			X
Tonga	X	X		X			X
Tuvalu	X	X	X	X			
Vanuatu	X	X	X	X	X		X

South America

Country/territory	CBD	UNFCCC	Kyoto Protocol	CCD	CITES	Ramsar Convention	World Heritage Convention
Argentina	X	X	X	X	X	X	X
Bolivia	X	X	X	X	X	X	X
Brazil	X	X	X	X	X	X	X
Chile	X	X	X	X	X	X	X
Colombia	X	X	X	X	X	X	X
Ecuador	X	X	X	X	X	X	X
Guyana	X	X	X	X	X		X
Paraguay	X	X	X	X	X	X	X
Peru	X	X	X	X	X	X	X
Suriname	X	X		X	X	X	X
Uruguay	X	X	X	X	X	X	X
Venezuela	X	X		X	X	X	X

Note: The regional breakdown reflects geographic rather than economic or political groupings.